Technical English 3

Workbook

Christopher Jacques

Pearson Education Limited

Edinburgh Gate
Harlow
Essex CM20 2JE
England

and Associated Companies throughout the world.

www.pearsonlongman.com

First published 2008
Seventh impression 2019

ISBN 978-1-4082-6798-1 (with key pack)
ISBN 978-1-4082-6799-8 (without key pack)

Set in Adobe Type Library fonts

Printed in Great Britain by Ashford Colour Press Ltd.

Acknowledgements

We would like to dedicate this book to the memory of David Riley, whose tireless professionalism contributed so much to its creation and success.

The publishers and author would like to thank the following for their invaluable feedback, comments and suggestions, all of which played an important part in the development of the course: Eleanor Kenny (College of the North Atlantic, Qatar), Julian Collinson, Daniel Zeytoun Millie and Terry Sutcliffe (all from the Higher Colleges of Technology, UAE), Dr Saleh Al-Busaidi (Sultan Qaboos University, Oman), Francis McNeice, (IFOROP, France), Michaela Müller (Germany), Małgorzata Ossowska-Neumann (Gdynia Maritime University, Poland), Gordon Kite (British Council, Italy), Wolfgang Ridder (VHS der Stadt Bielefeld, Germany), Stella Jehanno (Centre d'Etude des Langues/ Centre de Formation Supérieure d'Apprentis, Chambre de Commerce et d'Industrie de l'Indre, France) and Nick Jones (Germany).

Illustrated by HL Studios

The publisher would like to thank the following for their kind permission to reproduce their photographs:

(Key: b-bottom; c-centre; l-left; r-right; t-top)

Action Plus Sports Images: Mike Hewitt 59; **Alamy Images:** Alistair Laming 46 (G), Darrin Jenkins 46 (H), Jeremy Pembrey 46 (F), Keith Taylor 46 (E), Neil Grant 46 (B), Nigel Westwood 46 (C), Phil Degginger 30, Phil Wigglesworth 46 (D); **alveyandtowers.com:** 24; **Art Directors and TRIP Photo Library:** Helene Rogers 18, 54, 55r; **CastScope™ Tek84 Engineering Group, LLC., San Diego, CA:** 34t, 34b; **Corbis:** HYUNGWON KANG / Reuters 5; **FLPA Images of Nature:** Nigel Cattlin 63t; **Getty Images:** Hertfordshire Police 50, STR / AFP 62, Tobias Prasse 23; **Robert Harding World Imagery:** Worldscapes 45 (D); **iStockphoto:** 44t, 44b, 50b; **LMR Drilling UK Ltd:** 40l, 40r; **Reproduced with permission of Nissan:** 60l, 60r; **Photolibrary.com:** imageDJ 45 (B), NASA / The Print Collector 12; **Photoshot Holdings Limited:** Xinhua 45 (C); **Rail Images:** 49; **Restech Norway AS:** 19; **Rex Features:** 16, Action Press 52t, ITV 4, Sipa Press 39, Steve Hill 63b, Swani Gulshan 45 (A); **Science Photo Library Ltd:** Bernhard Edmaier 28, STARSEM 8; **Shutterstock.com:** iofoto 20; **STILL Pictures The Whole Earth Photo Library:** Biosphoto / Gilson François / BI 33; **SuperStock:** 46 (A); **Thinkstock:** Stockbyte 55l

Cover images: *Front:* **Alamy Images:** Technology and Industry Concepts

All other images © Pearson Education

Picture Research by: Kevin Brown

Every effort has been made to trace the copyright holders and we apologise in advance for any unintentional omissions. We would be pleased to insert the appropriate acknowledgement in any subsequent edition of this publication.

Designed by HL Studios, Long Hanborough

Cover design by Designers Collective

Contents

1 Systems

1 Rescue

1 Match the words and phrases 1–8 with the definitions a–h.

1 _h_ flares
2 ___ emergency beacon
3 ___ inflate
4 ___ satellite
5 ___ emergency signal
6 ___ winch (v)
7 ___ coastguard
8 ___ life raft

a) a series of radio waves that are sent in an emergency
b) a small rubber boat used by people from a sinking ship
c) to lift someone or something with a wire and a lifting machine
d) the organisation that helps boats in danger
e) a machine that is sent into space and orbits the earth
f) a device that sends a signal in an emergency
g) to fill something flexible with air so that it becomes larger
h) emergency devices that produce a bright flame

2 Use the words and phrases 1–8 in 1 to complete this news story.

Emergency beacon aids rescue from sinking boat ***23.12.09***

[1] Three men were rescued from a sinking fishing boat in the Gulf of Mexico today. The 32-foot-long boat was equipped with an (1) _emergency beacon_ **that** helped rescuers locate **the vessel** in the early-morning darkness, a Mexican Coastguard spokesperson said. The fishermen said they were asleep on the boat when a wave hit **their** vessel. They could not send a radio message or make a cell phone call.

[2] The (2) ______________ station in Veracruz was notified that the ground station had received an (3) ______________ from the boat's EPIRB (emergency position indicating radio beacon; this sends a signal **that** is picked up by a (4) ______________ and is transmitted to the ground station).

[3] A helicopter was despatched to the area. As **it** approached, one of the fishermen set off one of the red (5) ______________ **which** were kept on board, and the helicopter crew saw it.

[4] The fishermen were about to (6) ______________ their (7) ______________ when the helicopter reached **them**. The helicopter crew managed to (8) ______________ the three men to safety, and then flew them to the coastguard station, **where** they were given hot drinks and dry clothes.

3 Explain what the words in bold in 2 refer to.

1 that (para 1) _an emergency beacon_
2 the vessel (para 1) ______________
3 their (para 1) ______________
4 that (para 2) ______________
5 it (para 3) ______________
6 which (para 3) ______________
7 them (para 4) ______________
8 where (para 4) ______________

2 Transmission

1 02 Listen to this lecture about an FDR (flight data recorder). Underline the correct alternatives in the specification chart.

Operating frequency	*375 kilohertz / 37.5 kilohertz*
Maximum operating depth	*14,000 feet / 40,000 feet*
Frequency of transmission of signal	*once every 30 seconds / once per second*
Duration of signal	*13 days / 30 days*
Shelf-life of battery	*6 months / 6 years*
Method of transporting FDR if it has been in water	*in a container of ice / water*
Shape of beacon	*cylindrical / conical*
Colour of FDR	*orange / black*

2 Complete the description of how an FDR's locator beacon works. Use the correct form of one of the words in each pair for each gap.

activate/deactivate attach/detach manually/automatically receive/transmit release/fasten sink/float winch up / lower

The circular memory units with the flight data are stored in a large rigid cylinder that is (1) *fastened* onto the base of the FDR. The FDR is usually mounted in the tail section of the plane. In an accident, it becomes (2) ____________ from its mount. There is a submergence sensor on the side of the FDR's beacon. When water touches the sensor, this (3) ____________ the beacon (4) ____________. The beacon can (5) ____________ signals under water and above ground. Because of the weight of the FDR, it does not (6) ____________ on the surface of the water, but comes to rest on the seabed. After a diver has located the FDR on the seabed, it is (7) ____________ and transported to the computer lab for analysis.

3 Join these pairs of sentences into single sentences. Use *who*, *which*, *where*, *from where* to replace the words in italics.

Example: *1 ... TWA Flight 800, which crashed ...*

1 A serious air disaster occurred with TWA Flight 800. *It* crashed into the Atlantic in 1996.
2 The accident was caused by a build-up of fuel vapours in a fuel tank. *It* exploded.
3 Twelve minutes after take-off, the last radio transmission was received at Boston. *Here* the weather was fine.
4 The explosion was seen by another pilot. *He* was flying in the area at the time.
5 The other pilot landed at Boston airport. *From here* he contacted the air crash investigators.
6 An air and sea rescue was conducted in the area. *This* lies off the coast of New York State.
7 The FDR was recovered a week later by divers. *They* were guided to it by an emergency beacon.
8 The wreckage was transported to the shore. *From here* it was taken away for examination.

3 Operation

1 Describe the procedure for evacuating an aircraft. Complete the instructions, using the verbs in the box. Then match the instructions to the pictures.

> ensure fasten inflate place pull push release remove slide

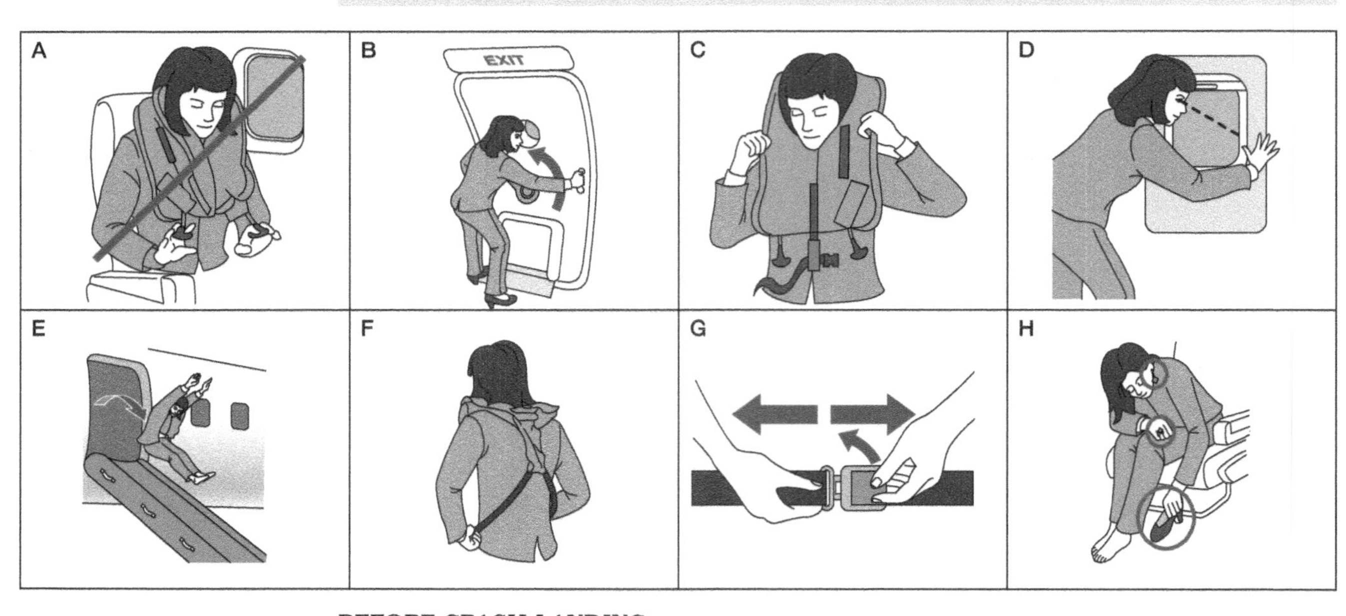

BEFORE CRASH-LANDING

1 ___Place___ your life vest over the head. _C_
2 ________ the straps around the waist. ___
3 Do not ________ the life vest while still in the aircraft. ___
4 ________ high heels and sharp objects. ___

AFTER CRASH-LANDING

5 ________ your seat belt. ___
6 ________ that there is no fire outside the emergency exit. ___
7 ________ down the red door lever, and ________ the door outwards. ___
8 When the slide is inflated, jump onto it and ________ down. ___

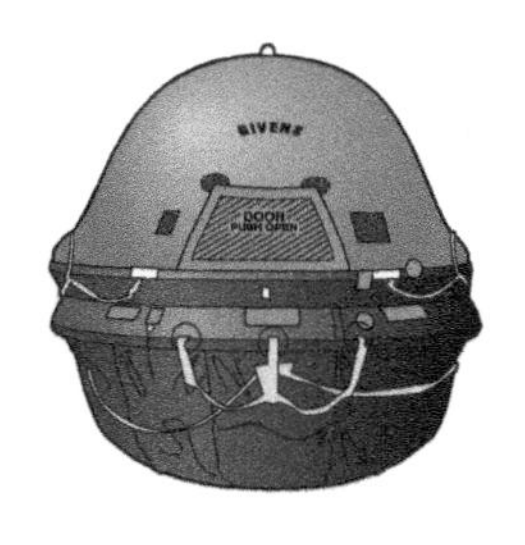

2 03 Listen to a demonstration of a life raft and underline the correct details.

1 Capacity: *4 / 8 / 12 people* 2 Inflation: *automatic / manual* 3 Storage: *in vinyl bag / in rigid container*	4 Stability: *can / cannot turn upside down* 5 Stability: *can / cannot right itself*

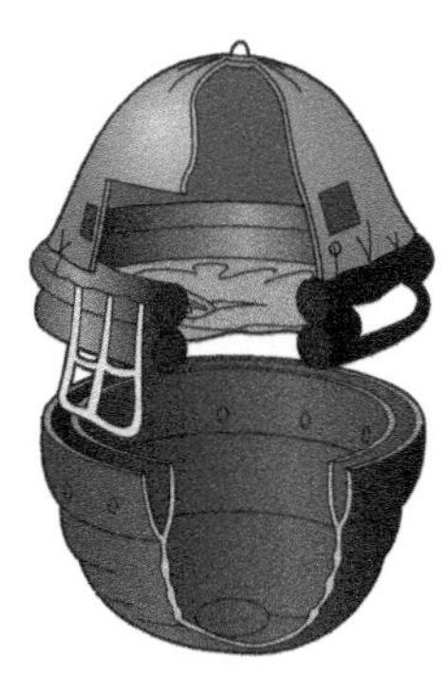

3 03 Listen again and complete the notes about the life raft.

1 **Construction:** ___inflated___ canopy (to keep out the water); ________ floor and canopy (to keep the occupants warm); water-________ lights on canopy; system for collecting ________
2 **Inflation:** inflation triggered ________; ________ buoyancy chambers inflated, to make the ________ of the raft; inflation forces ________ the carrying bag; inflation time: ________
3 **Stabilisation:** ________ chambers fill with water; upper chamber fills through portholes in the chamber ________; ________ chamber fills through a ________ valve, which ________ water in, but not out.
4 **Survival:** people in life rafts have survived hurricane conditions without capsizing (waves of > ________ metres, winds of > ________ kph). If capsizing occurs, life raft self-________ immediately.

4 Word list

NOUNS (rescue)	NOUNS (transmission)	VERBS	ADJECTIVES
beacon	antenna / antennas (plural)	activate	free-floating
coastguard		carry out	geostationary
flare	data (no plural)	convert	polar
life raft	emergency	detach	unseen
survivor	ground station	eject	visible
vessel	hydrostatic release unit (HRU)	ensure	**ADVERBS**
		float	automatically
	lever arm	inflate	manually
	magnet	locate	**PREPOSITIONAL PHRASE**
	megahertz	process	
	radio beacon	release	out of range
	tab	submerge	
	wavelength	winch	

1 Underline the one word which makes a compound noun with each word in bold.

1 **satellite** beacon signal wavelength
2 **low-altitude** centre orbit station
3 **radio** data power frequency
4 **operating** range length power
5 **air-sea** range beacon rescue
6 **safety** device antenna centre
7 **rescue** beacon team signal

2 Complete this paragraph, using the compound nouns in the box.

rescue centre ground station national centre emergency beacon
rescue team radio signal satellite signals

When the (1) *emergency beacon* switches itself on, a signal is sent to one or more satellites. A (2) __________ is then transmitted back to Earth to the (3) __________. The (4) __________ are processed at the ground station and converted into useful data. This data is then passed on to a (5) __________. The information about the location is forwarded to the nearest (6) __________, where a (7) __________ is sent out to look for the crashed ship or plane.

3 Label the pictures with nouns from the Word list.

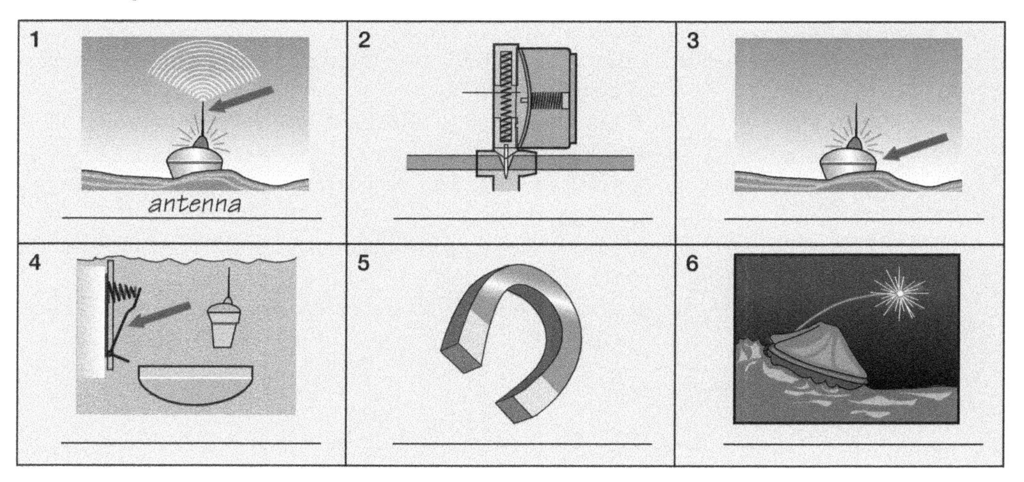

2 Processes

1 Future shapes

1 Complete this crossword.

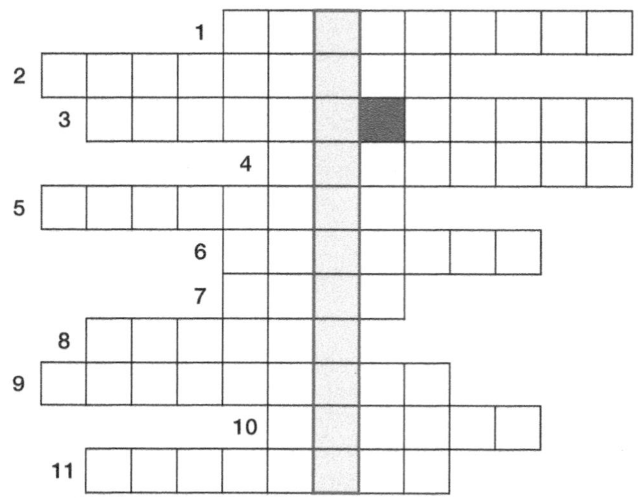

1 a mixture of two types of material
2 the industry that designs and builds planes and space vehicles
3 a plastic composite (2 words)
4 the main part or body of a plane
5 a plane or other vehicle that can fly
6 the material that is the topic of this unit
7 the horizontal part of a bridge
8 a written document on a particular subject
9 to build
10 a structure built over a river, road or rail track
11 a skilled person who designs, builds or maintains machines, engines, railways, etc.

Vertical word: to make

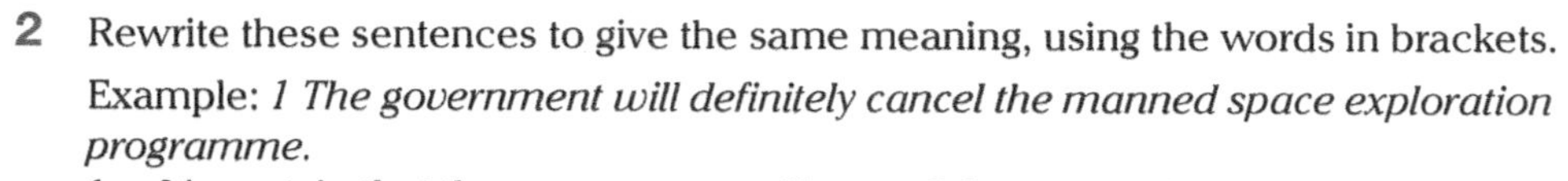

2 Rewrite these sentences to give the same meaning, using the words in brackets.

Example: *1 The government will definitely cancel the manned space exploration programme.*

1 It's certain that the government will cancel the manned space exploration programme. (definitely)
2 They will probably provide more money for robotic exploration of the solar system. (likely)
3 It's unlikely that scientists will develop new heavy-lift rockets in the near future. (probably)
4 It's definite that they will extend the life of the International Space Station beyond 2020. (certainly)
5 It's likely that they will ask commercial firms to play a bigger part in future. (probably)
6 It's possible that space travel to low-Earth orbit will become more affordable. (possibly)

2 Solid shapes

1 Match the parts 1–7 used during the injection moulding process with the descriptions a–g.

1 _f_ pellets
2 ___ melt
3 ___ heater
4 ___ rotation
5 ___ nozzle
6 ___ mould
7 ___ cavity

a) an electric- or steam-heated device which warms the cylinder
b) turning with a circular motion around a central point
c) to change from a solid to a liquid state
d) an empty space
e) a small hole through which soft plastic is pushed
f) small pieces of dry plastic used in the process
g) a hollow container in two halves used to shape a material

2 Read the text below about the process of rotational moulding. Put the pictures in the correct order. Complete the text, using the active or passive form of the verbs in brackets. One gap can be filled with the active or passive of the verb.

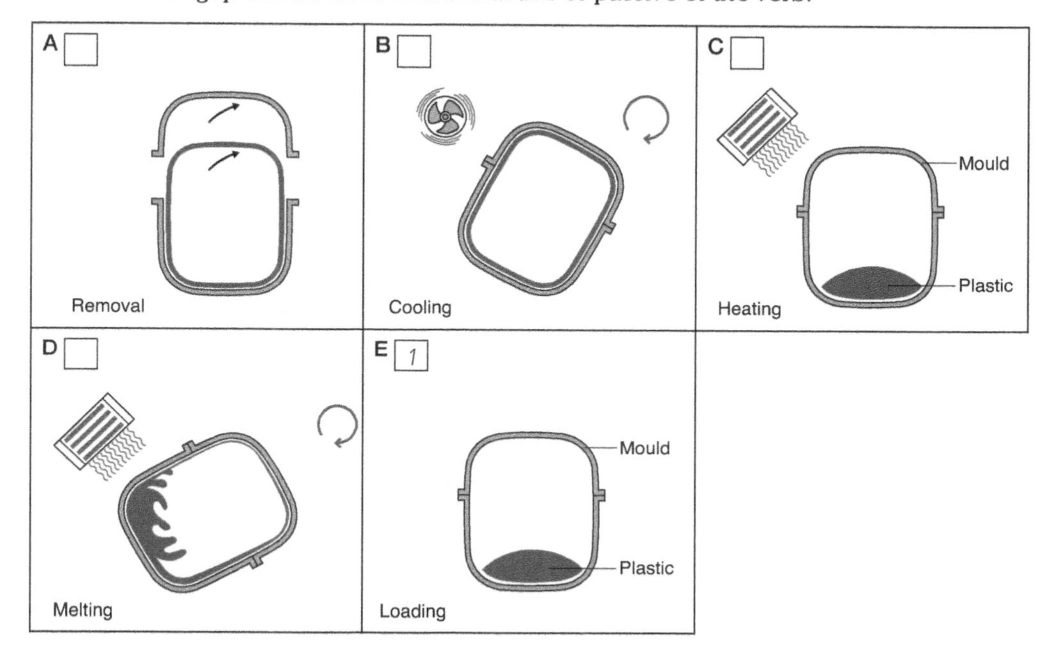

Rotational moulding is a high-temperature, low-pressure plastic forming process. It (1) _uses_ (use) heat and rotation (along two axes) to produce hollow, one-piece parts. The process is simple, but slow. It (2) ________ (use) for making large, hollow objects like oil tanks.

A quantity of plastic raw material, usually in powder form, (3) ________ (load) into the mould. The mould (4) ________ (heat) in an oven while it (5) ________ (rotate). The plastic raw material (6) ________ (melt) and (7) ________ (coat) the inside surface of the mould.

When the inside of the mould (8) ________ (coat), it (9) ________ (cool), usually by a fan. The cooling process takes between 10 and 20 minutes, depending on the size of the part and the thickness of the coating. As it (10) ________ (cool), the moulded object (11) ________ (shrink) and comes away from the sides of the mould. This makes it easier to take out the object at the end of the cooling process.

At the end, the object (12) ________ (remove). By this stage, the object is cool enough to handle.

3 Hollow shapes

1 Change these verbs to nouns and write them in the table.

blow cast close cool eject expand extrude heat
inflate melt move roll rotate transfer

-ing:	*blowing,* ______ ______ ______ ______	
-ion:	______ ______ ______ ______ ______	
-ment:	______	-er: ______
-ure:	______	no change: ______

2 04 Listen to this talk about vacuum forming. Tick the objects that are used in the process.

air pump ☐ hopper ☐ powder ☐
cylinder ☐ mould with holes ☐ solid mould ☐
heater ☐ pellets ☐ thermoplastic sheet ☐

3 Listen again. How does the speaker describe the four stages of the process? Write the names of the stages for each illustration. Label the parts a–d with the things that you ticked in 2.

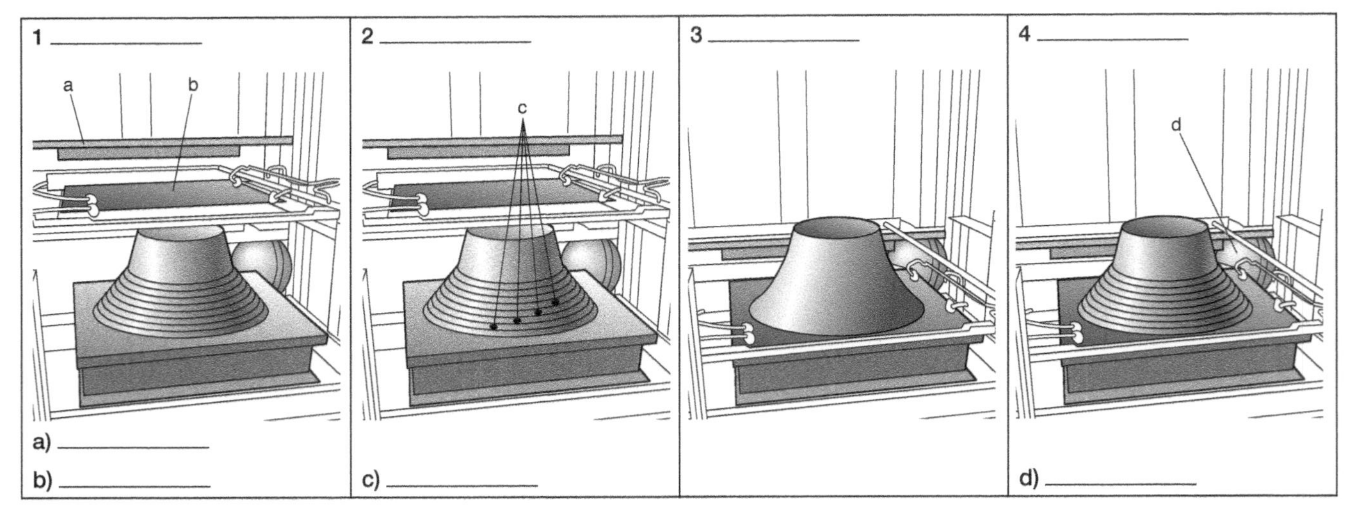

4 04 Listen to the talk again and complete the notes.

1 A thermoplastic sheet becomes *soft* with heating and ______ with cooling.
2 At the bottom of the vacuum mould, there are some ______-______ ______.
3 The sheet is positioned ______ the heater and ______ the mould.
4 When the sheet is stretched, it becomes ______.
5 The pump sucks out the air and this suction creates a ______.

4 Word list

VERBS (moulding)	NOUNS (moulding)	VERBS	COMPOUND NOUNS
close	closure	adjust	blow moulding
compress	compression	construct	injection moulding
eject	ejection	cool	metal-rolling
expand	expansion	design	pressure-die casting
extrude	extrusion	manufacture	**NOUNS**
heat	heater	melt	fibreglass
inflate	inflation	propel	fuselage
inject	injection	shape	**ADJECTIVES**
mould	mould	soften	molten
rotate	rotation		protective
transfer	transfer		specified
			unspecified
			PHRASE
			under pressure

1 Rewrite these sentences in the passive. Replace the active verbs in italics with the passive form of technical verbs from the Word list.

Example: *1 The polymer pellets are transferred from the hopper to the cylinder.*

Extrusion moulding

1 We *move* the polymer pellets from the hopper to the cylinder.
2 An electric motor *turns* the screw in the extrusion moulder.
3 The cold polymer pellets *move* along the cylinder.
4 Heaters *warm* the polymer pellets and turn them into a liquid.
5 The warm, soft molten polymer *moves* along the cylinder.
6 The machine *pushes* the molten polymer out into a mould.

Blow moulding

7 We *shut* the two halves of the mould with the molten polymer inside.
8 Compressed air *blows up* the molten polymer in the mould and makes it bigger.
9 The plastic bottle shape *gets colder*.
10 The machine *pushes out* the plastic bottle from the open mould.

2 05 Listen to and repeat the verbs and nouns in columns 1 and 2 of the Word list. Underline the main stress on the words of more than one syllable.

Examples: *closure*, *compress*.

3 Label the pictures with nouns from 1.

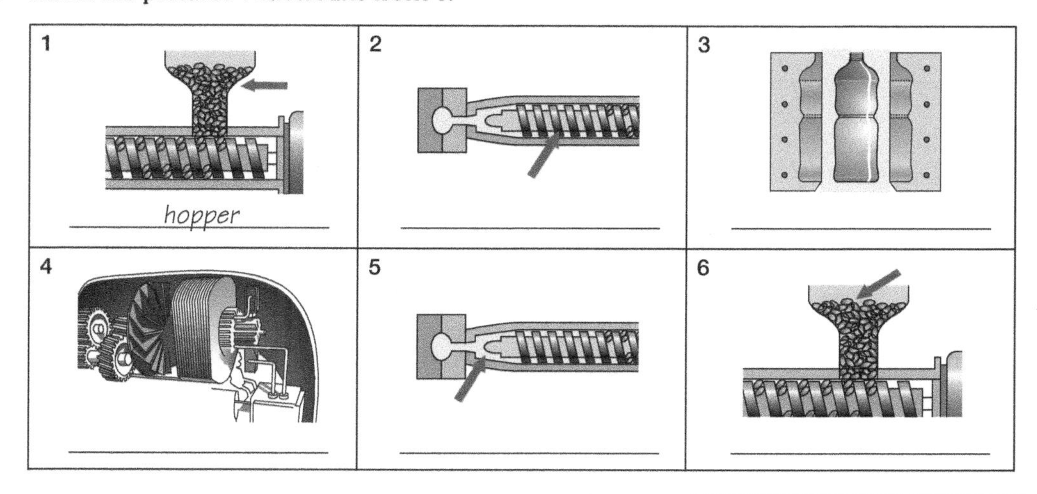

A Review

Section 1

1 Complete this report of a plane crash. Put the verbs in brackets into the past simple, and fill in each of the other gaps with one word.

On July 25th 2000, a Concorde aircraft (1) crashed (crash) near Paris. (2) It was travelling on a flight to New York. Emergency crews (3) ______ (race) to the area (4) ______ the crash happened to deal with the fire and search for survivors.

The Concorde (5) ______ down in flames just before 16.45 local time. Four hours after the accident, two Flight Data Recorders were found at the site of the crash. (6) ______ were taken to a laboratory for examination. The official accident report, (7) ______ was published in 2004, blamed the accident on a piece of metal, (8) ______ (9) ______ (fall) from another plane just before the Concorde (10) ______ (take) off.

The supersonic airliner (11) ______ (hit) this piece of metal and one of (12) ______ tyres (13) ______ (burst). As a result, pieces of rubber (14) ______ (fly) up and damaged a fuel tank. The fuel from the tank then (15) ______ (ignite), causing the fire and consequent crash.

2 Read some facts and figures about the Hubble Space Telescope. Match the information 1–12 with a–l.

1 k Launch date
2 ___ Launch vehicle
3 ___ Mass
4 ___ Type of orbit
5 ___ Orbit height
6 ___ Orbit time
7 ___ Orbit speed
8 ___ Telescope diameter
9 ___ Servicing Mission 4 date
10 ___ Batteries
11 ___ Number of radio antennas
12 ___ Goddard Space Flight Center (location)

a) near-circular low Earth orbit
b) May 2009
c) 8 km per second
d) Greenbelt, Maryland, USA
e) Space Shuttle Discovery
f) 11,110 kg
g) 6 × 57 kg nickel-hydrogen
h) 4
i) 97 minutes
j) 2.4 metres
k) 24th April 1990
l) 569 km

3 Complete this description of how the Hubble Space Telescope works, using the present simple active or passive of the verbs in the box.

avoid capture complete convert discover receive study transmit

Every 97 minutes, the Hubble Space Telescope (1) completes an orbit of the Earth. As it travels, its mirror (2) ______ light and directs it into its scientific instruments. Because the Earth's atmosphere partly blocks and distorts light, scientists (3) ______ these problems by placing their most advanced telescope 569 km above the Earth's surface, in a satellite.

The newest instrument on the satellite, Wide Field Camera 3, (4) ______ the formation of individual stars and (5) ______ new, extremely distant stars.

Instructions for the telescope (6) ______ from the ground station in the USA via a data relay satellite. These signals (7) ______ by antennas on the Hubble and (8) ______ into signals that point the telescope in the required direction and activate the instruments. Data goes from Hubble via the data relay satellite to the ground station.

Section 2

1 Which of these processes do the sentences below describe? Write the process letters A–D next to the sentences.

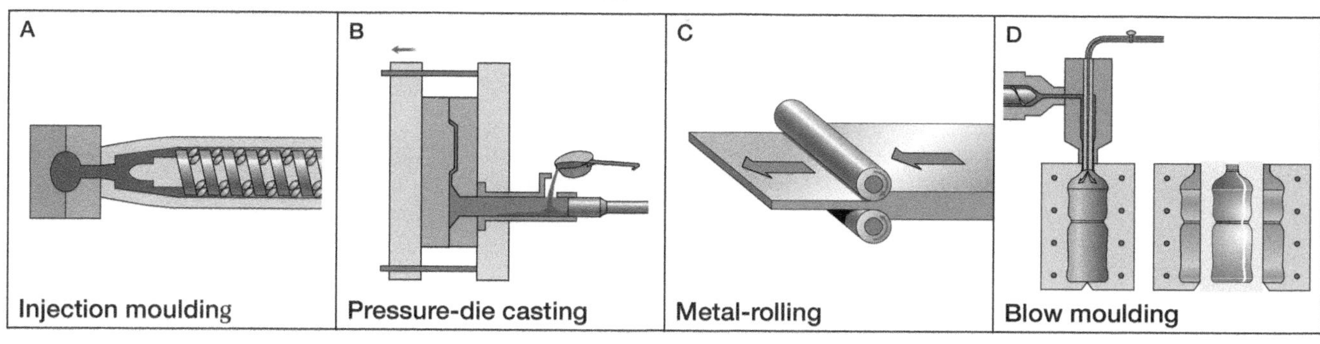

1 ___ The metal plate is first heated.
2 ___ The object, e.g. plastic bottle, cools and is then ejected from the mould.
3 _A_ The warm softened polymer is moved along the cylinder by a rotating screw.
4 ___ The molten metal is forced into the cavity between the two halves of the mould.
5 ___ Compressed air is blown into the molten polymer and inflates it.
6 ___ The mould opens and the hard, solid plastic object is ejected.
7 ___ After the metal cools, the mould is opened and the metal component is ejected.
8 ___ The metal sheet comes out from the rollers and is then cooled.
9 ___ The molten polymer expands to fit the mould, which may be in the shape of a bottle.
10 ___ The molten metal is pushed along the chamber under pressure by the injection piston.
11 ___ The metal plate is then pushed between the rollers, which compress it.
12 ___ The molten polymer is pushed by the ram through the nozzle into the mould, and becomes hard.

2 Write the three steps for each process in 1 in the correct order.

Injection moulding: ___ ___ ___ Metal-rolling: ___ ___ ___
Pressure-die casting: ___ ___ ___ Blow moulding: ___ ___ ___

3 Find the words in the puzzle and write them next to the definitions below.

P	O	L	Y	M	E	R	A	C	Y	F	G	X	W	J	Z	F	Y	W	P	I
R	O	L	L	E	R	U	Z	O	C	H	A	M	B	E	R	L	G	C	I	N
O	C	X	O	J	O	I	X	L	G	C	U	C	O	M	P	R	E	S	S	J
P	Z	I	F	E	X	P	A	N	S	I	O	N	I	Y	W	G	X	F	T	E
E	W	A	J	C	A	R	B	O	N	F	I	B	R	E	C	A	Z	U	O	C
L	S	O	F	T	E	N	F	K	U	Y	J	C	O	M	P	O	N	E	N	T

1 to make something soft _soften_
2 an increase in size ___________
3 a plastic composite ___________ ___________
4 a cylinder used to flatten a material ___________
5 a disc that is pushed along a cylinder ___________
6 an enclosed space inside a machine ___________
7 to throw something out of a mould or machine ___________
8 one of several parts that make up a machine ___________
9 to move something in one direction, e.g. along a cylinder ___________
10 to press something so that it takes up less space ___________
11 to push something inside through a small hole ___________
12 a chemical compound used for making plastics ___________

3 Events

1 Conditions

1 Complete these dialogues using the present perfect or past simple of the verbs in brackets.

1 A: You can't fly into the military airport. They (1) *have closed* (close) it.
B: Really? (2) ______ (there / be) an accident?
A: Yes. Two fighters (3) ______ (collide) on the runway in the fog last night.

2 C: Have you heard? One of our cargo ships (4) ______ (catch) fire in the Indian Ocean.
D: When (5) ______ (this / happen)?
C: Yesterday. The office (6) ______ (receive) a message early this morning.

3 E: How's the exploration going? (7) ______ (you / discover) any oil yet?
F: No, we (8) ______ (not / have) any luck yet.
E: How many wells (9) ______ (you / drill)?
F: (10) ______ (we / drill) three so far and now we're drilling the fourth.

2 Complete this article about space exploration. Use the second conditional of the verbs in brackets.

Are manned space missions to Mars a real possibility? At the moment, they don't seem likely and naturally we (1) *would only undertake* (only / undertake) missions if we (2) ______ (can / be) sure of getting the astronauts to Mars and back safely. But (3) ______ (the government / want) to provide money for the research if it (4) ______ (not / know) what the chances of success might be? And (5) ______ (it / choose) to spend millions on manned space missions to Mars even if there (6) ______ (be) plenty of money available?

On a manned mission, it (7) ______ (take) a long time for radio signals to reach the spacecraft from Earth, and so the crew (8) ______ (be) responsible for touchdown on the planet and for lift-off at the end of their research programme. There (9) ______ (be) no chance of a rescue mission if a disaster (10) ______ (strike) the crew while they were on Mars. And we (11) ______ (only / plan) further manned missions if our first mission (12) ______ (be) successful.

3 Write questions and answers to explain the procedure for take-off from an aircraft carrier. Use the first conditional.

1 Bad weather – what happens? Pilot cancels take-off
What will happen if the weather is bad? The pilot will cancel the take-off.

2 Launching device fails to function – what happens? Jet fighter remains on flight deck
______
______

3 Jet engine fails after take-off – pilot does what? Pilot activates ejection system
______
______

4 Pilot ejects after take-off – what happens? Parachute opens automatically
______
______

5 Pilot lands in sea after ejection – what happens? Helicopter winches pilot to safety
______
______

2 Sequence (1)

1 Read the text about the launch abort system in 6, Section 2 of the Course Book, page 23, and answer these questions.

1. Up to what altitude does the LAS operate?
   ____
2. In what situation is the LAS activated?
   ____
3. How is the LAS separated from the rocket?
   ____
4. How is the LAS controlled after it separates from the rocket?
   ____
5. With what kind of fixings is the crew capsule attached to the LAS?
   ____
6. What must happen in order to thrust the crew capsule away from the LAS?
   ____
7. What is the purpose of the parachutes of the crew capsule?
   ____
8. At what stage are the parachutes deployed?
   ____

2 Underline the correct linkers for describing the launch of the space shuttle.

(1) *Now* / *<u>Once</u>* the astronauts are in the space capsule, the access gantry is disengaged and swings away from the top of the rocket. (2) *Then* / *When* the fuel tank is filled with a mixture of liquid hydrogen and liquid oxygen. (3) *Afterwards* / *When* the countdown approaches zero, the ignition sequence is started. (4) *Then* / *As soon as* the rocket burners are ignited, the explosive bolts that hold the rocket in place on the launch pad are detonated and the rocket begins to lift off. (5) *After* / *While* ascending very slowly for the first ten seconds, the rocket accelerates. (6) *However* / *After* the three solid booster rockets have burnt out, they are jettisoned. Parachutes are deployed to slow their fall and they are recovered from the ocean and used again.
(7) *Then* / *However* the external fuel tank, which burns out later, is not recovered for reuse. It burns up as it falls through the Earth's atmosphere. (8) *Now* / *When* the spacecraft is stabilised and oriented by thrusters, and sent into orbit.

3 Describe the re-entry of the space shuttle, using the word(s) in brackets.

Example: *1 After giving the order for re-entry, the pilot fires the thrusters and turns the shuttle tail first.*

1. pilot gives order for re-entry → fires thrusters + turns the shuttle tail first (After giving)
2. shuttle reaches upper atmosphere → pilot fires thrusters again + turns the shuttle nose first (Once)
3. shuttle enters upper atmosphere → hot gases surrounding the shuttle cause radio blackout (When)
4. fully re-enters Earth's atmosphere → shuttle able to fly like an aircraft (As soon as)
5. picks up radio beacon at end of runway → pilot takes over control from the onboard computers (After)
6. pilot lands → deploys parachute from rear to slow the shuttle (After)
7. shuttle lands → crew follows procedure to shut down the shuttle (Once)
8. crew leaves shuttle → ground crew begin servicing it (As soon as)

3 Sequence (2)

1 Replace the more general verbs in italics with the correct form of the more technical verbs in the box.

deploy	detonate	eject	jettison	orient	propel	restrain	stabilise

1. Ships *are steadied* in rough seas by horizontal underwater bars. → are stabilised
2. A skydiver falls freely through the sky before *opening* a parachute. → ____________
3. Sailors and pilots can *find their position* by looking at the stars. → ____________ themselves
4. In an emergency, a safety system *throws out* the pilot through the top of the plane. → ____________
5. Plastic explosives are usually *set off* by an electrical current and a primer. → ____________
6. A jet plane is *pushed forward* by the backwards thrust from the jet engine. → ____________
7. Seat belts serve the purpose of *securing* drivers in their seats in the event of a crash. → ____________
8. If a ship is in danger of sinking, cargo can be *thrown off* from the side of the ship. → ____________

2 06 Listen to a talk about ejector seats and number these events in the correct order.

a) The pilot lands. ☐	e) The system ejects the pilot in his seat. ☐
b) The pilot tries to control the aircraft. 1	f) The pilot's main parachute opens. ☐
c) The drogue steadies the pilot's seat. ☐	g) The pilot pulls the ejection handle. ☐
d) The canopy over the cockpit is jettisoned. ☐	h) The pilot separates from his seat. ☐

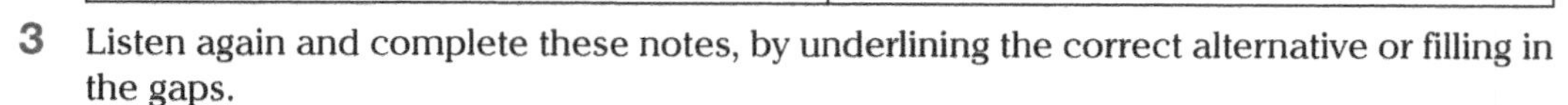

3 Listen again and complete these notes, by underlining the correct alternative or filling in the gaps.

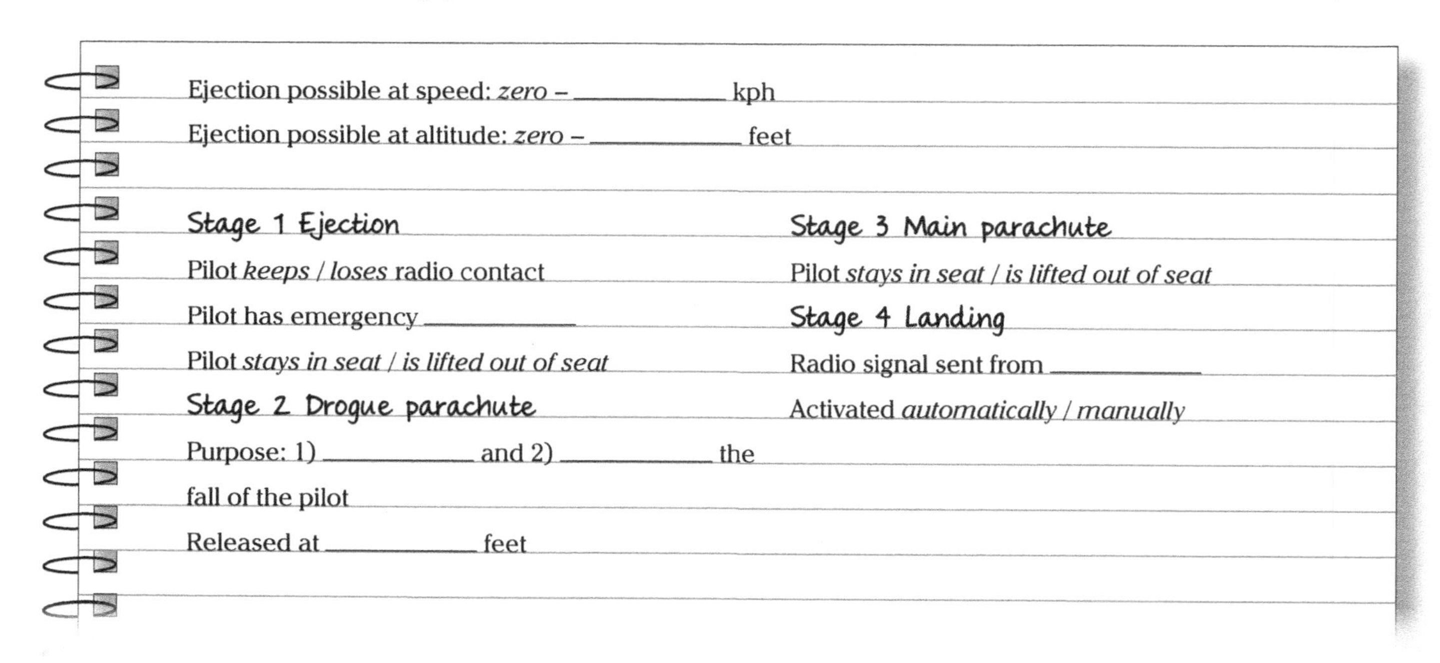

Ejection possible at speed: *zero* – ____________ kph
Ejection possible at altitude: *zero* – ____________ feet

Stage 1 Ejection
Pilot *keeps / loses* radio contact
Pilot has emergency ____________
Pilot *stays in seat / is lifted out of seat*

Stage 2 Drogue parachute
Purpose: 1) ____________ and 2) ____________ the fall of the pilot
Released at ____________ feet

Stage 3 Main parachute
Pilot *stays in seat / is lifted out of seat*

Stage 4 Landing
Radio signal sent from ____________
Activated *automatically / manually*

4 Word list

NOUNS	VERBS	PHRASAL VERBS	COMPOUND NOUNS
activation	activate	burn out	abort engine
astronaut	ascend	touch down	attitude-control engine
canopy	collapse	**ADJECTIVES**	
capsule	collide	explosive	crew capsule
catapult	deploy	medical	ejection system
cockpit	descend	professional	jet fighter
countdown	detect	world-class	jettison engine
deployment	detonate	**ADVERB**	launch abort system (LAS)
ejection	disable	automatically	
orientation	eject		oil rig
parachute	focus		
restraint	orient		
stabilisation	propel		
thrust	restrain		
	separate		
	spin		
	stabilise		
	thrust		

1 Complete these definitions with compound nouns from the Word list.

1. The function of an ______ oil rig ______ is to extract crude oil from the ground.
2. A ____________ is a type of military aircraft that attacks enemy aircraft.
3. The ____________ is the part of a spacecraft in which the crew travel.
4. The ____________ forms part of the nose, or tip of a rocket, and detaches itself if the rocket fails to launch successfully.
5. The function of the ____________ is to thrust the LAS, together with the crew capsule, away from the failed rocket.
6. The ____________ serves to stabilise the LAS (together with the crew capsule) and point it in the right direction.
7. The purpose of the ____________ is to push the crew capsule away from the LAS.
8. An ____________ serves to throw the pilot out of the cockpit if there is an emergency.

2 Check and write the past simple forms of these verbs from the unit.

break ______ broke ______

burn ____________ or ____________

fly ____________

focus ____________

propel ____________

sink ____________

spin ____________

strike ____________

take ____________

thrust ____________

3 07 Listen to and repeat the compound nouns in column 4 of the Word list.

4 Careers

1 Engineer

1 Read the blog and complete the profile.

My name's Martha Bari and I live in Ottawa, Canada. I'm Italian and I've lived in Ottawa for seven years. I came to Ottawa to study at the university. In my fourth year, I got my Masters in Bio-medical Engineering, after doing a degree in Mechanical Engineering for three years.

I love my work. I speak French as well as Italian, which is useful in Canada. Right now, I'm working for a bio-medical company in Ottawa called Robotico. I'm a Research & Development Technician and I'm developing new devices for artificial arms and legs. My ambition is to become the Head of R&D, but don't tell my boss that!

By the way, are any other graduates from my university out there (University of Ottawa, 2004–2008, particularly anyone from the Department of Mechanical Engineering)? Please post a message.

1	Name	*Martha Bari*
2	Nationality	
3	Employer	
4	Current position	
5	Responsibilities	
6	Qualifications	
7	Institution	
8	Skills and competences	

2 Complete these blogs with the verbs in brackets, using the present simple, present continuous, or *going to* future. There may be more than one possible answer.

1

My name's Kees. Right now I (1) *am working* (work) at a manufacturing company in Rotterdam. The company (2) ______ (specialise) in robotics. I normally (3) ______ (spend) four days a week with my employer, and one day at college. At the moment I (4) ______ (do) an apprenticeship in engineering. After I (5) ______ (complete) my apprenticeship in July, I (6) ______ (have) a short holiday, probably somewhere in the sun! I (7) ______ (not have) any further career plans at present.

2

My name's Pedro and I (1) ______ (work) for a pharmaceutical company in Madrid. Most days I (2) ______ (work) from 8.30 until 17.00, but on some days I (3) ______ (stay) late in order to finish a job. At the moment I (4) ______ (develop) a new line of over-the-counter drugs. We (5) ______ (start) our first series of trials in two months' time, so there's a lot of last-minute work. I already (6) ______ (have) a Bachelor of Sciences degree in Biochemistry, and I (7) ______ (start) a Masters course next autumn.

2 Inventor

1 Complete the description of a line-thrower, using the words in the box.

barrel bullet device hand-held pulse recoil (x2) roughly

A pneumatic line-thrower is a (1) ___*device*___ used for throwing a long line, either between a pair of boats, or from a boat to the shore. It is also used for rescue purposes. It is a (2) ______ device, (3) ______ 75 cm in length, and uses compressed air as a propellant.

The standard pneumatic line-thrower consists of a reservoir of compressed air, a long (4) ______ and a trigger. The sudden release of a (5) ______ of compressed air propels the projectile, which may be shaped like a ball or a (6) ______. When fired, the pneumatic line-thrower causes a (7) ______ in the opposite direction to the line of fire: the bigger the device, the greater the (8) ______. The line is stored in a separate box and follows the projectile when it is fired.

2 Write questions about the MoorLine 230.

		MoorLine 230	Line-thrower 75
1	Line length	240 m	100 m
2	Line thickness	3.2 mm	5 mm
3	Line breaking strength	2000 N (newtons)	1500 N
4	Projectile	cylindrical projectile	plastic ball
5	Range	230 m	90 m
6	Recoil	5400 N (newtons) (max)	5200 N (max)

1 How ___*long is the line?*___
2 How ______
3 How ______
4 What ______
5 How ______
6 What ______

3 Write sentences comparing the two products in 2, using these prompts.

1 line / 230 / long / 75
The line on the 230 model is longer than the one on the 75 model.
2 line / 230 / thick / 75
______
3 breaking strength / 230 / while / 75
______
4 230 / have / projectile / whereas / 75 / ball
______
5 75 / short / range / 230
______
6 230 / great / recoil / 75
______

3 Interview

1 ▶ 08 Read part of a CV and then listen to an interview. Complete the missing sections and update some of the information.

CURRICULUM VITAE	
PERSONAL INFORMATION	
Surname / First name(s)	GALLINI, Laura
Position applied for	Technician ______
WORK EXPERIENCE	
Dates	2005–______
Position	Junior Technician
Responsibilities	______
Name and address of employer	Horton Engineering, Cleveland
Type of business	______
WORK EXPERIENCE	
Dates	______ – ______
Position	______
Responsibilities	Machining, finishing and some ______
Name and address of employer	Farley Marine, Long Creek
Type of business	Marine engineering: manufacture of engines and pumps
EDUCATION AND TRAINING	
Dates	______ – ______
Qualification	______
Subjects / Skills covered	______ and ______
Name of institution	Albany College of Engineering, Albany

2 ▶ 09 Write the questions from the interview, using the prompts. Then listen again and check.

1 which / job / interested *Which job are you interested in?*
2 long / work / there ______
3 responsibilities / current / job ______
4 exactly / kind / business / your company ______
5 long / work / Farley Marine ______
6 what / description / there ______
7 long / course / last ______
8 why / leave / Horton Engineering ______

3 Write the time words and phrases in the box on the correct lines.

> 2011 three weeks yesterday seven years a week last week January six months 8 o'clock Monday a month two hours 20 minutes 12th May five days lunchtime a long time

for: *three weeks,* ______

______

since: *2011,* ______

______

4 Word list

NOUNS	NOUNS	ADJECTIVES	NOUNS (careers)
accuracy	prototype	accurate	ambition
barrel	pulse	external	apprentice
blog page	recoil	genetic	apprenticeship
brain cell	robotics	high-pressure	benefit
brass	software	inaccurate	bonus
breakthrough	spacer	interpersonal	competence
bullet	target	minimum	institution
cell	tissue	optimum	role
DNA	**VERBS**	organisational	technician
gene gun	insert	pharmaceutical	**NOUNS (qualifications)**
helium	modify	previous	
inaccuracy	power	social	Bachelor of Sciences
leisure activity	**ADVERB**	standard	certificate
machine gun	roughly		degree
mechatronics			diploma
membrane			Masters

1 Complete this description of a gene gun, using the correct form of words from the Word list.

A (1) *machine gun* is a weapon that fires (2) ________ down a cylindrical (3) ________ towards a (4) ________. Holes in its barrel allow air to escape and this reduces the weapon's (5) ________, i.e. the thrust of the weapon in the opposite direction to the line of fire.

(6) ________ are tools which fire DNA into a cell. They are powered by (7) ________ gas. A (8) ________ of the gas pushes the (9) ________ bullet through the cell (10) ________ and into the (11) ________ itself.

The first (12) ________ gene gun was unsatisfactory. It was too (13) ________ and caused (14) ________ damage to the cell. The (15) ________ gene gun worked much better. It had a narrower barrel and fired the DNA with greater (16) ________.

2 10 Write the words in the box in the correct column. Then listen and check.

ambition apprentice apprenticeship benefit certificate competence diploma engineering institution interpersonal previous technician

1st syllable stressed	2nd syllable stressed	3rd syllable stressed
	ambition	

B Review

Section 1

1 Identify which five paragraphs below relate to take-off from an aircraft carrier, and which five relate to landing. Then put the paragraphs in the correct order.
Take-off: ___ ___ ___ ___ _a_
Landing: ___ ___ ___ ___ ___

a) If everything goes well, the aircraft has enough lift to get airborne. However, if there is a mishap, the pilot can activate the ejector seat to exit the cockpit before the plane crashes into the sea.
b) After this, the plane is pulled off the landing strip and fastened securely with metal chains.
c) At this stage, the plane does not move, because the shuttle is restrained by a holdback bar. Finally, the catapult officer releases the holdback system and the catapults propel the fighter plane forwards. At the end of the catapult run, the tow bar is released automatically, allowing the plane to rise into the air away from the flight deck.
d) The catapults consist of a small **shuttle**, which is attached to two pistons. These pistons are inside two parallel cylinders that run below the flight deck. When preparing for flight, the **tow bar** on the plane's nose gear is attached to the shuttle.
e) To come to a stop on a short flight deck, an aircraft must decelerate very quickly.
f) At the end of the countdown, the catapult officer opens the valves, which fill the catapult cylinders with high-pressure steam. This steam will provide sufficient pressure to propel the shuttle forwards. At the same time, the pilot increases the jet engines to maximum thrust.
g) One final detail: as soon as the plane strikes the flight deck, the pilot increases the engine thrust to full power. As a result, if the plane does not engage one of the arresting wires, it will have enough thrust to lift off into the air again at the end of the flight deck.
h) For this reason, it has a **tail hook**, which is an extended hook attached to the tail. When approaching the flight deck, the pilot must engage the tail hook with one of four **arresting wires**. These are steel cables that are stretched across the width of the **flight deck**.
i) To get airborne from an aircraft carrier, a jet fighter needs extra acceleration. This is provided by **catapults**, which are fixed into the flight deck.
j) If everything goes perfectly, the pilot is guided along the correct descent path by **coloured lights** on the aircraft carrier. He also visually lines up the **runway centre line** painted on the flight deck, and the **vertical drop line** painted on the ship's stern below the end of the flight deck. When he touches down, the tail hook engages with one of the arresting wires connected to a hydraulic cylinder. The hydraulic system slows the plane and brings it to a standstill within 100 metres.

2 Use the words printed in bold in 1 to label the diagrams.

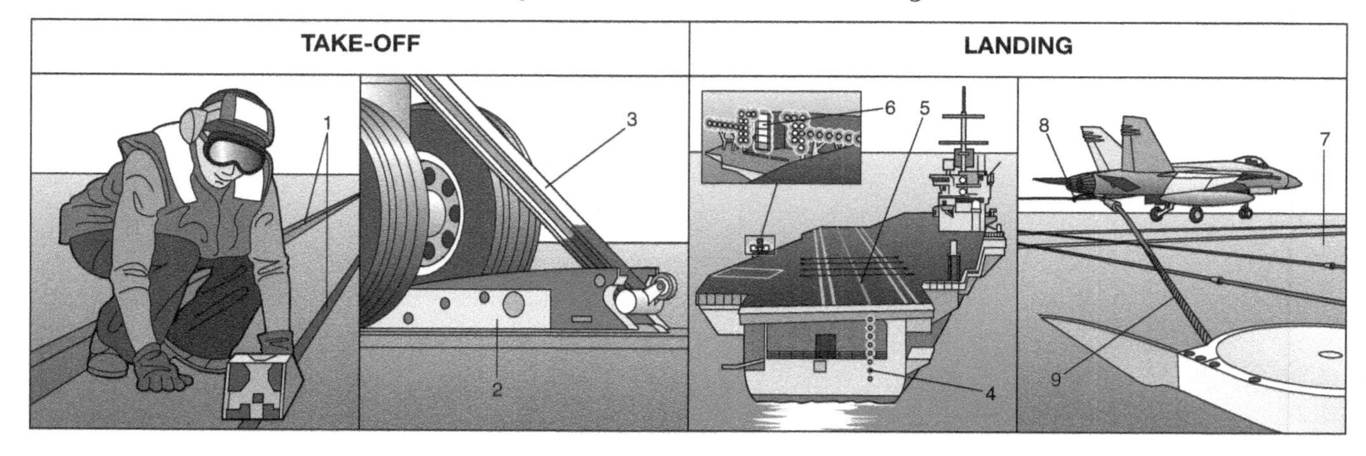

Section 2

1 Make compound nouns using the words in the box. There may be more than one possible answer.

> activity car degree description experience page plan qualification technician vitae

1 apprentice *technician*
2 blog ______
3 business ______
4 career ______
5 company ______
6 curriculum ______
7 engineering ______
8 job ______
9 leisure ______
10 work ______

2 Complete this dialogue between two HR managers. Use all the words in the box or the correct form of the words in brackets where given.

> for (x2) since (x2) ago but (x3) while whereas

A: What did you think of those two candidates, Enrico and Manuel?
B: Well, Enrico seemed (1) *more confident* (confident) than Manuel.
A: Yes, (2) ______ Manuel is much (3) ______ (old) and (4) ______ (experienced).
B: True, but I think Enrico would pick up the job (5) ______ (fast) than Manuel. He asked lots of good questions and showed (6) ______ (great) knowledge of our company.
A: Let's look at their relative work experience. Enrico (7) ______ (work) in his first company (8) ______ three years and (9) ______ (work) in his present company (10) ______ the last four years.
B: But Manuel (11) ______ (be) out of work (12) ______ January. He (13) ______ (leave) his last job ten months (14) ______ and he (15) ______ (not work) (16) ______ then. I wonder why.
A: He said he had been ill, (17) ______ you can't be sure. By the way, (18) ______ (you / receive) their references?
B: Yes, Enrico has (19) ______ (positive) references than Manuel.
A: So, would you consider Enrico to be the (20) ______ (good) candidate?
B: Yes, he showed more ambition. Anyway, let's look at their qualifications.
A: Enrico did a four-year apprenticeship, (21) ______ Manuel only did three years. I wonder why.
B: Enrico has a degree in Mechanical Engineering, (22) ______ Manuel only has a diploma. (23) ______ (you / check) their documentation?
A: Yes, Manuel studied for the diploma, (24) ______ he left before the end of the course and didn't take the final exams.
B: So his qualifications are not just (25) ______ (impressive), they're non-existent!

5 Safety

1 Warnings

1 ▶ 11 Listen to five phone conversations between a customer and a manufacturer's helpline. Number the problems in the order in which you hear them.

No.	Problem	Remedial action required
	a) oil	continue to monitor / change oil / top up with oil
1	b) ABS	stop driving and seek immediate assistance / consult main dealer
	c) fuel	empty contaminated fuel from tank / fill up with fuel
	d) tyres	check tyre sensors / deflate tyre / inflate tyre
	e) doors	check all doors are shut / check electrical contact / replace warning bulb

2 Listen again and underline the remedial action required for the problems in 1.

3 ▶ 12 Complete the questions and answers from 1. Then listen again and check.

1 'How is the car handling? Do you need to take ___any corrective action___ when you brake?'
'Yes. When I ______, I have to ______.'

2 '______ is the warning light and ______?'
'It's in the middle of the ______ and it's ______.'

3 'Is it on ______?'
'No, I've been ______ all day. It just comes on ______.'

4 'And what ______ is ______?'
'I'm getting ______.'

5 'Is it a ______, or both?'
'There's a ______ when I turn on the ______.'

4 ▶ 13 Underline the most suitable discussion markers in this conversation about safety devices. Then listen and check.

A I sent you a proposal about a new in-car warning system that would warn drivers when they get too close to the vehicle in front. (1) *Anyway* / *For example*, this could be a tactile or audible warning.

B (2) *In other words* / *By the way*, you're suggesting a beep or something that makes the steering wheel vibrate.

C (3) *I don't think so* / *I agree with that.* I think we're on the right lines here. (4) *Alternatively* / *By the way,* talking of lines, I was late today because of a pile-up in the fog. I think a lorry had gone off the road and …

A Yes, yes. (5) *By the way* / *Anyway,* you're here now, so let's get back to the subject. (6) *In other words* / *Alternatively,* we could have a visual warning sign, (7) *or* / *for instance* one that was flashed onto the inside of the driver's windscreen.

B (8) *I don't think that's a good idea* / *You have a point.* There are too many flashing lights already when you're driving.

A Well, what about a system based on different sensors, to monitor things like the distance from the vehicle in front, your speed, the road conditions and the external temperature? Then we're combining different factors.

C (9) *I don't think that's a good idea* / *That sounds good.* (10) *So* / *Or,* an audible or tactile warning system related to different factors.

2 Instructions

1 Complete the descriptions of drum brakes and a handbrake, using the labels in the diagrams. One label is used twice.

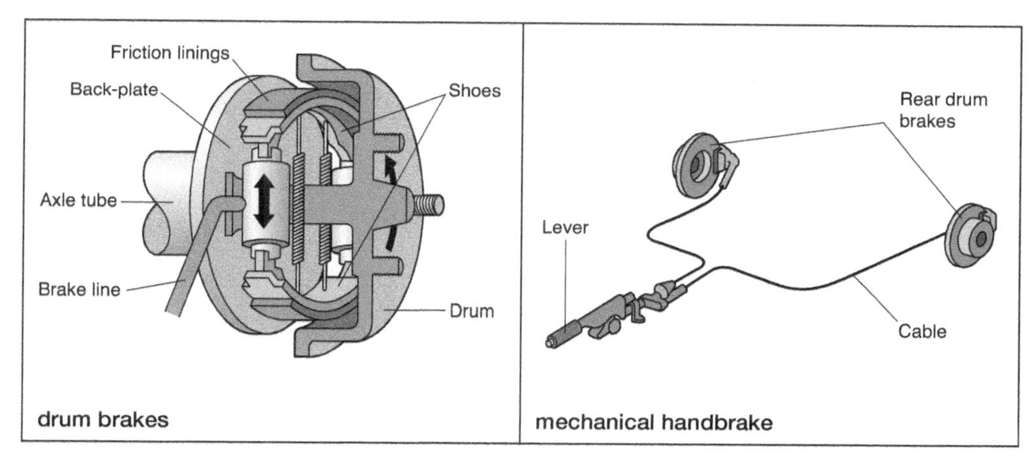

Drum brakes consist of two (1) ____shoes____, which are fixed to a (2) ________________, which is fixed to an (3) ________________. (4) ________________ are riveted to the outer faces of each shoe. When the brake pedal is depressed, this increases the hydraulic pressure in the (5) ________________ that runs from the master cylinder. This forces the two (6) ________________ into contact with a rotating (7) ________________, which is fixed to the wheel hub by the wheel nuts. The inner surface of the drum is ground smooth, so that the shoe linings can rub against it.

A **handbrake** must be fitted to every car. This holds the vehicle stationary while it is left unattended. It can also function as an emergency brake if there is a failure of the main braking system. Normally, the handbrake operates on the (8) ________________ and is linked to them via a (9) ________________. The handbrake mechanism is operated by a (10) ________________, which is held in the 'on' position by a ratchet and pawl mechanism.

2 Write sentences for a car manual using the checklist below. Use *should*, *need to* and *don't have to* with the verbs in the box, and the active or passive as appropriate. Some verbs are used more than once.

adjust change check refill replace top up

Examples: *1 The oil and the filter need to be changed. …*
4 You need to check the screen wash system.

Jobs for garage	Jobs for you
1 Necessary: oil and filter change; radiator – anti-freeze; windscreen replacement (if badly cracked)	4 Necessary: screen wash system; headlamps (if going to a country where people drive on a different side of the road)
2 A good idea: air bag replacement after ten years	
3 Not necessary: windscreen replacement (if in good condition)	5 A good idea: oil level; tyres (pressure / tread depth); screen wash; lights check (side lights, headlights, indicators)
	6 Not necessary (with improved technology): battery check; radiator top-up

3 Rules

1 Read this news story about a near miss and label the diagram with the names of the two aircraft.

Air traffic control procedures have been changed after a near miss by two aircraft at a military airfield last June.

The incident involved a Hawk training jet and a twin-engine turboprop Reims-Cessna aircraft. The Hawk was about to land when the Cessna flew across the runway at a very low altitude. The air traffic controller in the control tower had given the pilot of the Cessna permission to fly north-north-east low over the airfield shortly before the incident.

The air traffic controller was dealing with a high workload at the time, and was controlling the movements of five different aircraft on four different radio frequencies.

A report by the Civil Aviation Authority concluded that the controller was unable to handle this large workload, thus endangering aircraft in the area. There had also been a breakdown in the monitoring of flight levels around the airfield, and of radio communications.

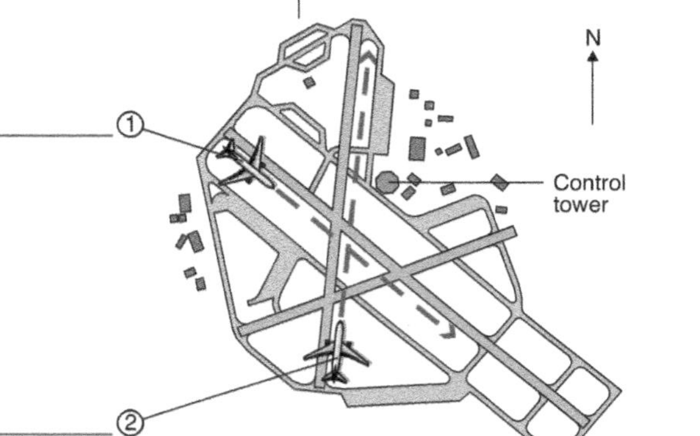

Neither pilot was informed that other aircraft were flying in the same area, the report added. Neither pilot had time to take evasive action, and a collision was avoided by pure luck. At their closest distance, the two aircraft were about 150 metres apart.

Staffing rotas for air traffic controllers at the airfield have now been modified to ensure that this does not happen again.

2 Are these statements about the text in 1 *true* (T) or *false* (F)? Correct the false ones.

1 Air traffic control procedures at the airfield are likely to be modified in future.
2 The turboprop aircraft had permission to cross the airfield at low altitude.
3 The danger of a collision was caused by poor communications alone.
4 The air traffic controller intervened to warn the pilots about other aircraft.
5 New staffing rotas are likely to endanger aircraft at the airfield in future.

3 Rewrite these instructions to give a similar meaning, using the words in brackets. Do not use the words in italics.

Example: *1. Pilots who are unfamiliar with the layout of this airport must order a 'Follow-Me' car before they have landed.*

ARRIVALS

1 Pilots who are unfamiliar with the layout of this airport *should not* order a 'Follow-Me' car *after* they land. (must / before)
2 *You should only* ask for priority landing *if* you have mechanical problems, a medical emergency or are dangerously short of fuel. (not / unless)
3 Do *not* order mobility buggies and wheelchairs *before* you have landed. (only / after)
4 Pilots *should only* shut down the engines *when* the ground staff have indicated that the aircraft is correctly positioned at the gate. (must not / until)

DEPARTURES

5 *You must only* embark onto your plane *after* filing a flight plan, with details of your route and destination. (not / before)
6 *Do not* take off *before* ensuring that you have sufficient fuel for the flight + 10%. (only / after)
7 In snowy weather, aircraft *should only* attempt take-off *after* de-icing in the holding area. (must not / without)
8 *Do not* move onto the runway *unless* you have received permission from the control tower. (only / if)

4 Word list

NOUNS (cars)	VERBS (driving)	VERBS	ADJECTIVES
brake line	accelerate	devise	apart
brake pad	counter-steer	drain	audible
brake pedal	cruise	endanger	corrective
cable	depart	ensure	evasive
calliper	drift	evacuate	even-numbered
disc (brake)	overtake	grind	mandatory
drum (brake)	**NOUNS**	intervene	odd-numbered
fluid level	altitude	maintain	prohibited
handbrake	collision	modify	remedial
hydraulic fluid	flight level	monitor	spongy
indicator	heading	process	tactile
master cylinder	helicopter	regain	visible
piston	high-visibility clothing	scrape	**ADVERBS**
reservoir		squeal	intentionally
sensor	image		unintentionally
steering wheel	incident		
torque	maintenance		
windscreen	naked flame		
	softness		
	vibration		

1 Label this diagram with words and phrases from the Word list.

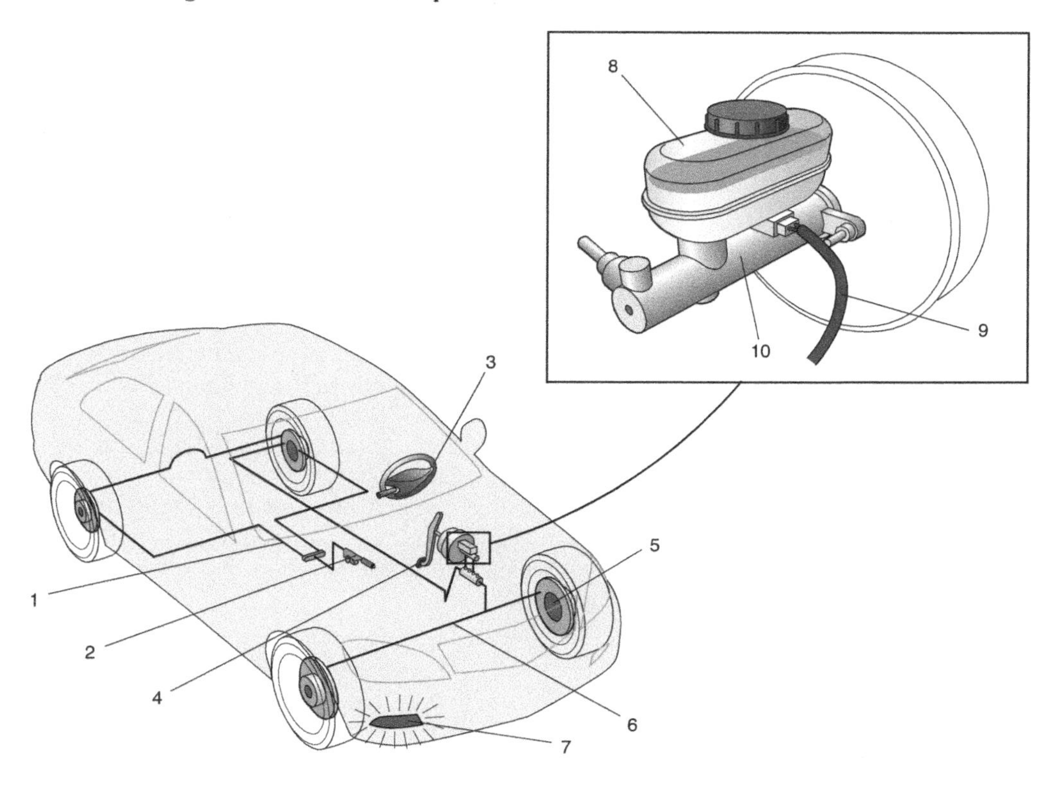

6 Planning

1 Schedules

1 ▶ 14 Listen to part of an interview about climate change and energy sources. Complete the table.

Predictions and targets	Change in °C or %	By date
Carbon emissions to cause rise in global temperatures	a) + _0.5_ °C	b) _already_
(75% possibility)	c) + ______ °C	d) by _2060_
(50% possibility)	e) + ______ °C	f) by ______
Need to stabilise carbon emissions Need to reduce carbon emissions	g) + _0_ % i) by ______ %	h) by ______ j) after ______
Percentage of energy output from non-carbon-based fuels	k) ______ %	l) by ______
European targets: reduce carbon emissions	m) by ______ % o) by ______ %	n) by _2020_ p) by ______

2 ▶ 15 Complete the questions, using the phrases in the box. Then listen to the interview again and check your answers.

will that reduce could we avoid will push up you think in the long term your views could happen

1 Do _you think_ that carbon emissions ______ global temperatures?
2 Is this the worst that ______ in the future?
3 So how ______ this?
4 What are ______ on switching energy sources?
5 So ______ carbon emissions ______?

3 Complete these sentences of agreement or disagreement.

1 A: I think we should restrict all travel that uses carbon-based fuels, including flying.
B: I don't _agree with you_ at all. I just can't go ______ that.

2 A: There are cost and environmental benefits in switching to electrically powered vehicles.
B: You have ______ there. I'm happy ______.

3 A: CCS is surely the way forward, both in the short term and medium term.
B: I'm not ______ that. Let's think again about it.

4 A: Can we agree that we can achieve a 10 percent reduction in carbon-based fuels next year?
B: That ______ right. Yes, that's ______ me.

2 Causes

1 Rewrite each sentence to give a similar meaning, making the following changes.

- replace *because* with the phrase in brackets. Example: *because* → *owing to*
- replace the verbs in italics with related nouns. Example: *reacts* → *reaction*
- replace the adverbs in italics with adjectives. Example: *completely* → *complete*

Example: *1 Phosphorus (used in the manufacture of detergents) is stored in water owing to the reaction of phosphorus with air.*

1. Phosphorus (used in the manufacture of detergents) is stored in water *because* phosphorus *reacts* with air. (owing to)
2. The reaction happens *because* calcium carbonate *is added*, together with water, to the gas. (due to)
3. The gas is purified *because* polluting particles *are completely removed* from the gas by the collection plates. (as a result of)
4. Emissions from power plants are lower with clean coal technology *because* the coal *is totally purified* before it is burnt. (as a result of)
5. Port installation costs are high *because* the LNG *is liquefied* at a temperature of −1,620 Celsius before it is loaded onboard. (owing to)
6. Greater crop yields are ensured *because* the growing sheds *are automatically humidified.* (as a result of)
7. The emergency rescue teams had problems *because* the volcanic ash *rapidly solidified* after cooling. (caused by)
8. The coal used in the power plant is a fine dust *because it is pulverised* in the coal mill nearby. (due to)

2 Complete this crossword with words from Section 2 of the Course Book, pages 44–45.

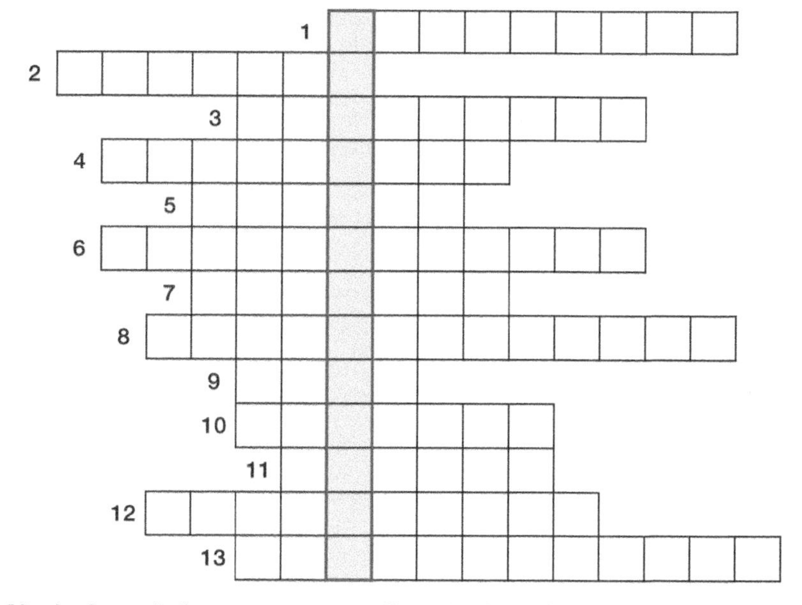

Vertical word: these remove small particulates from flue gas

1. very small pieces of matter
2. a common element (S) found in carbon-based fuels
3. a piece of metal that causes particles to become negatively charged
4. a general word for substances used in or obtained by chemistry
5. containing salt
6. to remove or separate sulphur from something
7. a layer of rock that can hold water
8. an ___ dust precipitator is a type of dust filter
9. a mistake, mark or weakness that makes something imperfect
10. a colourless, odourless, flammable gas (CH_4)
11. to form ions, or make them form
12. substances in another substance that make it of poor quality
13. the process of converting into a gas

3 Systems

1 Match the signpost phrases a–j with the notes for a presentation about geothermal energy in Iceland.

a) OK, let's move on to production wells.

b) I'd like to conclude by showing this slide.

c) Now let's look at transmission pipelines.

d) Thanks, Ed. Now let's look at generating electricity.

e) Now I'd like to move on to the power plant.

f) The aim of my talk is to tell you about …

g) Thank you all for coming. Any questions?

h) Now I'd like to hand over to Ed, who will look into the question of district heating.

i) Next, let's look at different field types.

j) Let's start with the underground rock layers.

	Phrase	Presentation notes
1	*f*	geothermal power in Iceland: electricity generation, space heating, etc.
2		hot rock layers from a volcanic system; fissures join together to create an underground reservoir; becomes an aquifer of hot water, extracted through production wells
3		low-temperature fields: yield water at below 150 °C for space heating high-temperature fields: yield water at above 200 °C for electricity generation and heating cold water for space heating
4		13 production wells at Nesjavellir, in operation since 1972, to supply the capital Reykjavik: thermal production for space heating, with generation of electricity
5		**show slide:** transmission pipeline. Distance from Nesjavellir to Reykjavik: 27 km. Pipeline diameter: 90 cm. Water temperature: 96 °C. Temperature loss in transit: 1–2 °C
6		first stage of the process: central separation station: steam is piped through moisture separators to steam heat exchangers inside the building
7		steam from the production wells can't be used for space heating; problem of mineral-rich saline solution which blocks pipes. Instead, heat exchangers are used to heat cold water, which is then distributed
8		power plant capacity: 120 MW of electrical power (started production 1990)
9		**show PP slide:** Statistics: Iceland's energy requirements geothermal energy: 24.5% hydro-electric: 75.4% fossil fuels: 0.1% geothermal energy: provides heating and hot water for 87% of buildings
10		invite audience to ask questions

2 Answer these questions, using the notes in 1.

Example: *1 Because hot water can be extracted from the underground reservoirs.*

1 Why is Iceland such a good location for geothermal energy?
2 Where does Reykjavik get some of its geothermal energy from?
3 How does the hot water get from Nesjavellir to Reykjavik?
4 Why isn't steam from the extraction wells used directly for space heating?
5 How does Iceland get most of its energy requirements?

4 Word list

NOUNS	NOUNS (energy)	VERBS (energy)	COMPOUND NOUNS
deadline	flaw	condense	bio fuel
timeline	gasification	consume	carbon capture and storage (CCS)
NOUNS (chemicals)	geothermal energy	convert	
	humidification	desulphurise	
calcium carbonate	impurity	emit	clean coal technology
carbon dioxide (CO_2)	injection	gasify	
methane (CH_4)	insertion	humidify	collection plate
nitrous oxide (N_2O)	ionisation	ionise	flue gas
ozone (O_3)	liquefaction	liquefy	iron filing
sulphur	magnet particle	pressurise	power plant
NOUNS (energy)	particulate	pulverise	**ADJECTIVES**
	precipitator	purify	electrostatic
aquifer	pulverisation	recover	harmless
desulphurisation	purification	solidify	saline
electrode	reaction	sulphurise	volcanic
emission	reservoir	switch	**VERBS**
extraction	solidification		finalise
fissure	sulphurisation		obtain
			scan

1 Match 1–8 with a–h to describe causes and effects.

1 _e_ Electrostatic precipitators remove
2 ___ CO_2 can be stored
3 ___ Methane can be recovered
4 ___ Pressure from the CO_2
5 ___ A chemical reaction results in
6 ___ Polluting particles are attracted
7 ___ The collection plates get
8 ___ The iron filings move because

a) the desulphurisation of the flue gas.
b) from underground coalfields.
c) by the collection plates.
d) in underground saline aquifers.
e) small particles from flue gas.
f) the magnet attracts them.
g) forces the oil to rise to the surface.
h) a negative electric charge from the electrodes.

2 Complete the sentences with the correct form of energy verbs from the Word list.

1 To stop power plants ___emitting___ harmful gases, they should be ______________ to CCS.
2 To make coal cleaner, you can ______________ it into small particles and ______________ it before burning.
3 After going through the separator, the steam ______________ and is then returned to the reservoir.
4 Gas can ______________ and ______________ to LNG by cooling it to a very low temperature.

3 16 Listen to and repeat the verbs in column 3 of the Word list, and the compound nouns in column 4.

C Review

Section 1

1 Match each language function 1–9 with one example in column 2 (a–i) and another in column 3 (j–r).

Function	Examples (a–i)	Examples (j–r)
1 agree d, q 2 give an example ___ ___ 3 approve of an idea ___ ___ 4 change the subject ___ ___ 5 give another possibility ___ ___ 6 disagree ___ ___ 7 suggest ___ ___ 8 go back to the main subject ___ ___ 9 say the same thing differently ___ ___	a) in other words, b) Alternatively, c) I think we should use … d) I agree with that. e) I don't think that's a good idea. f) I like that. g) But to return to the subject, h) By the way, i) for instance,	j) I can't go along with that. k) That sounds good. l) that is, m) Anyway, n) Why don't we have a …? o) such as p) To change the subject for a moment, q) You have a point there. r) Or, to consider a different possibility,

2 Match these signs to the prompts below.

1 staff / not / enter / building (A)
2 doors / lock / night (P)
3 fork-lift trucks / not / drive / this part / warehouse (P)
4 workers / not / eat / drink / this area (A)
5 all visitors / report / reception (A)
6 pedestrians / not / walk / red zone (A)
7 machine guards / use / all times (P)
8 ear protection / wear / when / machinery / operation (P)

3 Use the prompts in 2 to write a sentence about each of the signs. Use *must* or *must not* and the active (A) or passive (P).

Example: *1 Staff must not enter the building.*

Section 2

1 Complete the sentences for a presentation about incineration, using prepositions.

1 The aim ___of___ my talk this morning is to explain ______ incineration.
2 First ______ all, let's discuss the problem of waste.
3 Now I'd like to move ______ ______ the main elements of the incinerator.
4 So now I'm going to talk you ______ the process of modern incineration.
5 Now I'd like to hand ______ ______ Jan.
6 Jan will look ______ the question of pollution.
7 Thanks, Jan. Finally, I'd like to conclude ______ showing this slide.
8 Thank you all ______ coming. Any questions?

2 Read this email enquiry. Write a reply, using the presentation notes below to help you answer the questions.

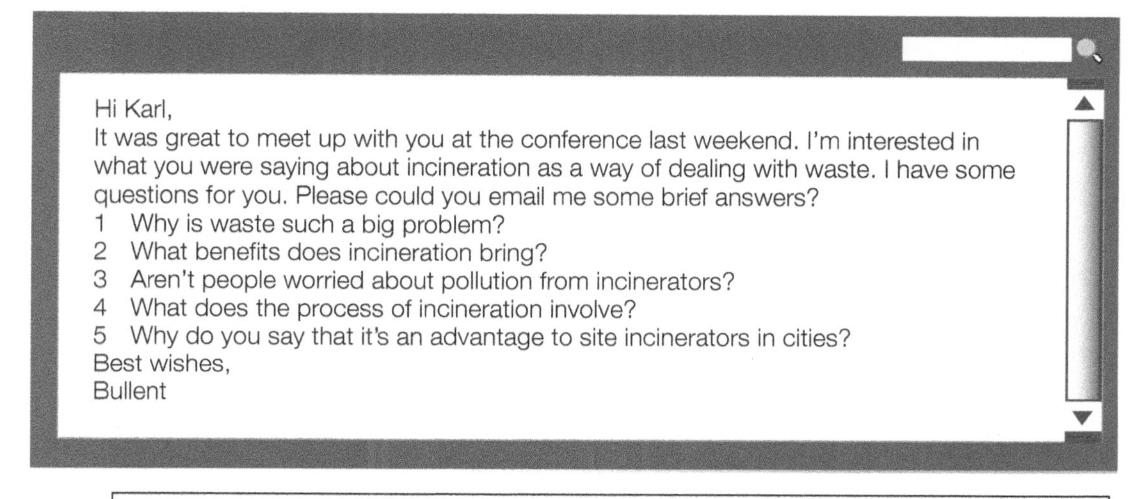

Hi Karl,
It was great to meet up with you at the conference last weekend. I'm interested in what you were saying about incineration as a way of dealing with waste. I have some questions for you. Please could you email me some brief answers?

1 Why is waste such a big problem?
2 What benefits does incineration bring?
3 Aren't people worried about pollution from incinerators?
4 What does the process of incineration involve?
5 Why do you say that it's an advantage to site incinerators in cities?

Best wishes,
Bullent

Presentation notes
Incineration converts waste materials → bottom ash, flue gases, particulates and heat. Heat used for space heating and/or electricity generation.
Waste is bulky; incineration reduces mass of waste by 80–85%.
Show slide: moving grate incinerator. Continuous process: waste enters at one end; ash leaves at the other end. Heat transformed into steam → drives the turbine. Flue gases are cooled → pass through the flue gas cleaning system.
Bottom ash from the incinerator – non-hazardous. People worried about pollution from incinerators. Improved emission control: pollutants in the flue gas reduced by particle filtration, using electrostatic precipitators. Heavy metals, e.g. mercury, lead, chromium, removed by acid gas scrubbers.
Show slide: Incinerator in city centre. Emission control better in modern incinerators. Convenient for reception of waste, and for output of space heating and electricity. Can increase electricity generation in summer or space heating in winter.

Hi Bullent,
Many thanks for your email asking about incinerators. I'll try to answer …

7 Reports

1 Statements

1 Underline the correct phrases. There may be more than one possible answer.

1. I said *the official* / *to the official* / *(NO WORD)* that I had a metal plate in my hip.
2. He told *me* / *to me* / *(NO WORD)* the airport had put in new security procedures.
3. I informed *the clerk* / *to the clerk* / *(NO WORD)* that I had packed my case myself.
4. She reported *me* / *to me* / *(NO WORD)* that there had been a security breach.
5. She explained *him* / *to him* / *(NO WORD)* that the X-ray machine didn't harm laptops.
6. She promised *him* / *to him* / *(NO WORD)* she would phone as soon as she arrived.
7. He confirmed *me* / *to me* / *(NO WORD)* that he had orders to carry out a manual search.
8. He assured *them* / *to them* / *(NO WORD)* that there would be a full investigation.

2 Complete the table below with the reporting verbs in the box.

say tell inform report explain promise confirm assure

1 verb + object	*tell* ________ ________
2 verb + *to* + object OR without object	________ ________ ________ ________
3 verb + object OR without object	________

3 Write what the security inspector actually said, using direct speech.

Example: *1 'The airport has introduced a new scanning system.'*

1. The security inspector reported that the airport had introduced a new scanning system.
____
2. He said the device was able to produce X-ray images of plaster casts.
____
3. He confirmed that it would be installed at all airports in the following six months.
____
4. He informed us that criminals had been able to transport knives inside plaster casts.
____
5. The security inspector told his assistant to conceal a knife inside the demonstration cast.
____
6. He said the body part being examined was placed next to the device, not in it.
____
7. He explained that the device uses a new system called 'backscatter imaging'.
____
8. He assured us that backscatter imaging could be used safely on all passengers.
____
9. He told us that the first group of operators was already undergoing training.
____
10. He ordered us to read the user manual before the following day.
____

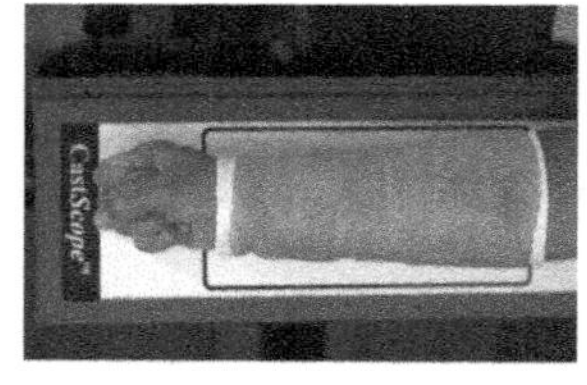

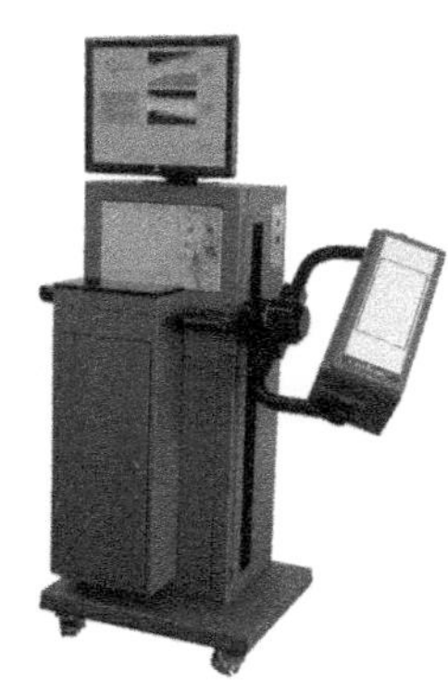

2 Incidents

1 Match the words and phrases in the box with their definitions 1–8.

audio circuit coil detection zone direct current magnetic field pulse-induction technology reflected pulse resistor

1 a horizontal space in which objects can be detected ___detection zone___
2 an area around an object that has magnetic power ______
3 a short electrical current caused by the collapse of the magnetic field in the coil ______
4 a flow of electricity that moves in one direction only ______
5 a wire arranged in a continuous spiral ______
6 an electrical circuit which creates an audible sound ______
7 a wire or object that stops or slows down the flow of electric current through it ______
8 a system which uses a coil or coils as both transmitter and receiver ______

2 Read the product review in 2, Section 2 of the Course Book, page 54. Are these statements *true* (T) or *false* (F)? Correct the false ones.

1 The metal detector can operate at a distance of 65 m from the passenger's body.
2 The display screen shows where a metal object is located on the operator.
3 The WTMD is based upon eight linked coils of wire, arranged horizontally.
4 An electrical current is sent through each of the eight coils in pulses.
5 The presence of a metal object creates a magnetic field around it.
6 A reflected pulse disappears faster after a metal object is detected.
7 A metal object is present on a passenger if a reflected pulse lasts longer than expected.
8 The audio output sounds deeper and quieter as the metal object approaches the coil.

3 Underline the correct alternative in these sentences. There may be more than one possible answer.

1 *As / While* the pilot was preparing to land, a mobile signal interfered with the radio.
2 The plane was flying *while / when* the mobile phone was seeking the nearest ground station.
3 The pilot approached the runway *while / as* the mobile phone was causing interference.
4 *While / When* the plane was waiting on the runway, a warning signal sounded.
5 Another plane was taking off *when / as* the stick-shaker signal in the cockpit started to operate.
6 *When / While* the Boeing 737 was flying, a passenger was using his mobile phone.

4 Use these notes to make sentences similar to those in 3.

Example: *1 While the passengers were checking in, the ground crew was refuelling the Airbus.*

1 07.35–07.52: passengers check in / 07.35–07.52: ground crew refuels Airbus
2 07.55–08.15: ground crew loads luggage into Airbus / 08.10: starts to snow
3 08.27: 747 lands / 8.20–8.31: ground crew de-ices Airbus
4 08.31: 747 slides off runway / 08.28–08.31: 747 taxis towards terminal building
5 08.15–08.45: Airbus passengers wait in departure lounge / 08.40: hear that flight is cancelled
6 08.55–09.15: Airbus passengers ask for information / 747 passengers were waiting to leave their aircraft

3 Progress

1 ▶ 17 Listen to a dialogue about iris scanning. Number the topics in the order in which you hear them.

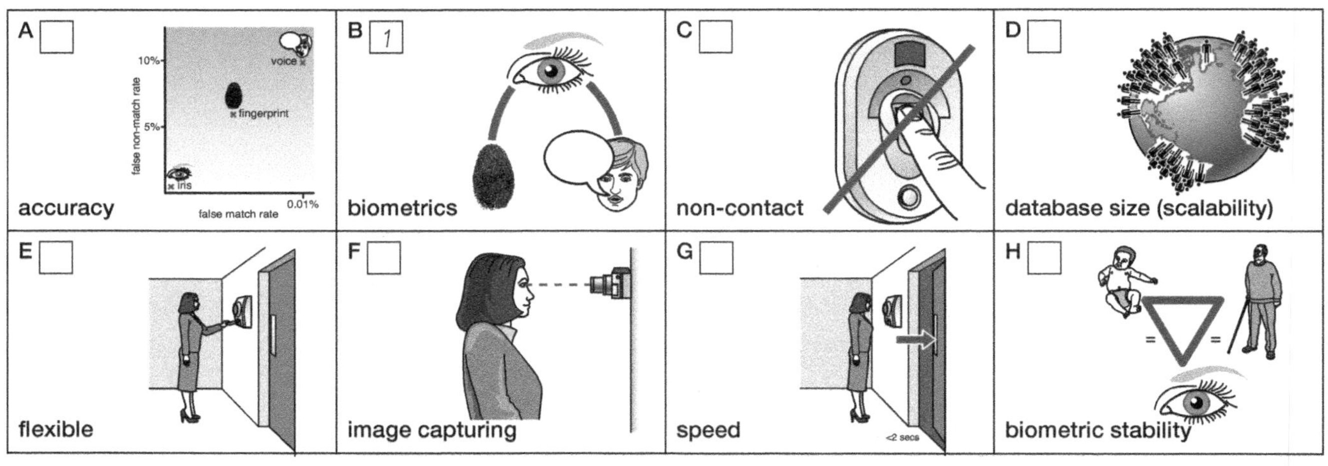

2 ▶ 18 Put the words in the correct order to make sentences from the dialogue. Then listen again and check your answers.

1 you on how getting are? *How are you getting on?*
2 you have systems with up what come? ______
3 what system we that about that other about spoke? ______
4 that have looked into you? ______
5 you see what I mean yes ______
6 more on me go tell ______
7 the how so operate system does? ______
8 you catch after I'll weekend with the up ______

3 Complete this text, using the words and phrases in the box.

8 and 35 cm   activated   converted   scanning   digital   iris identity
PIN numbers   security door   video camera technology   database

Iris recognition technology provides accurate identity security without (1) *PIN numbers*, passwords or cards. Verification takes less than two seconds. Although the term 'iris scanning' is commonly used, there is no (2) ______, in fact. The methodology is based on (3) ______. For registration, the person is positioned between (4) ______ from the auto-focus camera. Then a (5) ______ video is taken of the iris. Individual images are taken from the video. The patterns of the iris are (6) ______ into a 512-byte digital template. This is stored in a (7) ______ and is accessed by the Identification Control Unit. When a person approaches a Remote Optical Unit (ROU), it is (8) ______ by proximity sensors. The ROU uses the same video and image capture technology to compare the present-time image with the (9) ______ stored at the time of registration. As soon as the iris is matched, a direct signal is sent to open a (10) ______, or otherwise confirm the person's identity.

4 Word list

NOUNS (electronics)	NOUNS	ADJECTIVES	COMPOUND NOUNS
amplifier	breach	hand-held	baggage X-ray machine
capacitance	criminal	innocent	
capacitor	fingerprint	manual	capacitive scanning
capacitor plate	investigation	non-conductive	
charge circuit	jaw	**VERBS**	CCTV camera
coil	official	affect	detection zone
conductor	password	beep	fingerprint scanning
conductor plate	pin number	convict	
dielectric	procedure	detect	hand-held metal detector (HHMD)
discharge circuit	ridge	discharge	
magnetic field	security	disguise	iris scanning
magnetism	source	scan	optical scanning
microchip	surgeon		pulse-induction
reflected pulse	surgery		remote display unit
resistor	valley		voice recognition
sensor			walk-through metal detector (WTMD)
terminal			

1 Complete the descriptions of these diagrams with the words in italics in column 1 of the Word list. One word is used twice.

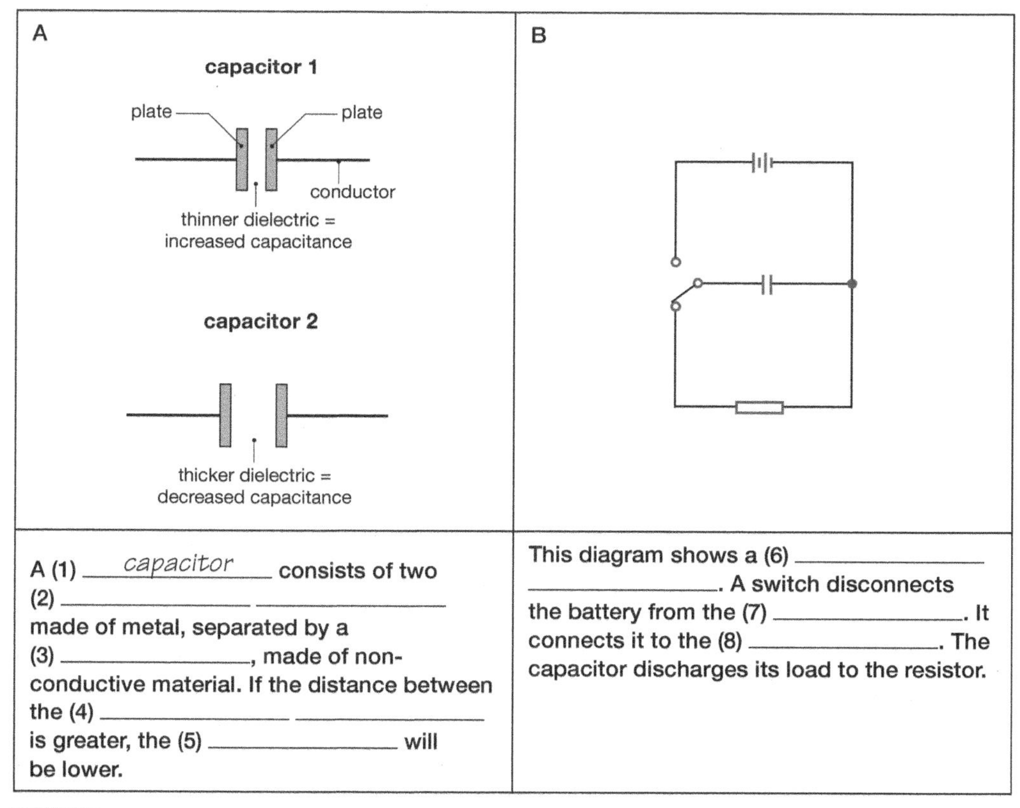

A A (1) capacitor consists of two (2) ______ ______ made of metal, separated by a (3) ______, made of non-conductive material. If the distance between the (4) ______ ______ is greater, the (5) ______ ______ will be lower.

B This diagram shows a (6) ______ ______. A switch disconnects the battery from the (7) ______. It connects it to the (8) ______. The capacitor discharges its load to the resistor.

2 19 Listen to and repeat the highlighted nouns in column 1. Underline the syllable with the main stress in each word.

8 Projects

1 Spar

1 Label the parts of this illustration using the words in the box.

crude oil pipeline   power plant   pipeline network   mooring line
pumping station   seabed   tank farm   tug boat (for towing tankers)

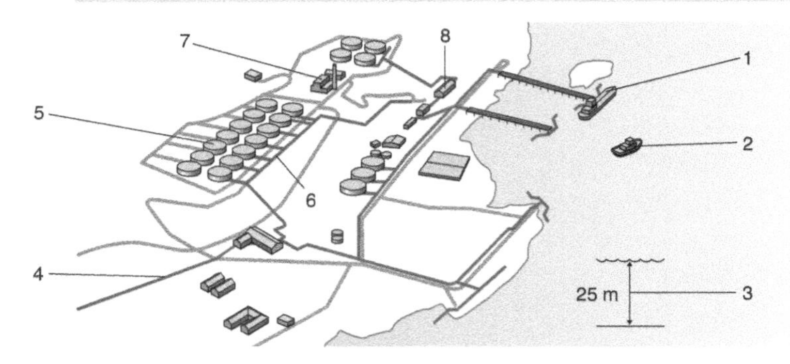

2 Look at the GANTT chart for a fire detection and suppression project at a new marine terminal. Write short dialogues, using *already*, *ago*, *yet* and *not yet*. It is now the end of the third week of February.

Example:

1 *A: Have the foundations for the pumping station been laid yet?*
B: Yes, they were laid in January, four weeks ago.
A: What about the pumps? Have they been installed?
B: No. That job has already been started, but it hasn't been finished yet.

		J	A	N		F	E	B		M	A	R	
		w	e	e	k	w	e	e	k	w	e	e	k
	Task	1	2	3	4	1	2	3	4	1	2	3	4
	Pumping station												
1	Lay foundations	■	■	■									
	Install pumps						■	■	■				
2	Construct site roads	■	■	■	■								
	Install pipeline ring main						■	■	■	■			
	Water storage tanks												
3	Drill test boreholes		■										
	Lay foundations				■	■							
4	Order tank sections			■									
	Assemble tank sections									■	■	■	
5	Deliver fire guns						■						
	Use fire guns in training								■		■		■

2 Platform

1 20 Listen to a recorded talk at the Visitors Centre of the Three Gorges Project (TGP) in China. Tick the visitors' questions that are answered in the talk.

1 Did people use to die because of flooding in the area? ☑
2 Does the Yangtze River still flood? ☐
3 How many power plants are there? ☐
4 In what year was the TGP completed? ☐
5 Did some of the displaced population decide to move to other areas? ☐
6 Has any new housing been provided for the displaced population? ☐
7 When was the project started? ☐
8 Could the dam be in danger if there was an earthquake in the area? ☐

2 Listen again and answer the questions in 1 that you ticked.

3 21 Listen to further details about the TGP and complete the table.

Earth and rock excavation	a) 103 million m^3	Construction period	e) ______ 2009
Earth and rock fill	b) ______ m^3	People employed on site	f) ______
Concrete	c) ______ tonnes	Reservoir area	g) ______ km^2
Rebar	d) ______ tonnes	Reservoir length	h) ______ km

4 Complete the sentences using *by means of*, *to* or *by* and the correct form of the verb in brackets.

1 The Yangtze River was diverted during the project to allow (allow) dam construction to proceed.
2 The first stage of river diversion was achieved ______ (construct) a 550 m long embankment.
3 The embankment was constructed ______ (dump) rock into water 60 m deep.
4 Shipping up to 10,000 tonnes can proceed upstream or downstream ______ a 5-step ship-lock.
5 A ship elevator was constructed ______ (pass) smaller vessels (up to 3,000 tonnes) through the dam more quickly.
6 Electricity is generated ______ hydro turbines and generator units.
7 22 spillway openings are installed along the top of the dam ______ (allow) extra discharge outlets.
8 The reservoir level is lowered during May and June each year ______ (provide) extra storage during the flood season from June to September.

3 Drilling

1 Match the sentence halves 1–7 with a–g, and 8–14 with h–n. Pay particular attention to the words in bold.

1 _f_ The **derrick** is	a) on the **drilling platform.**
2 ___ The **derrick** stands	b) the **cable,** and moves up or down.
3 ___ The **crown block** is fixed	c) or releases the **cable.**
4 ___ The **winch** pulls in	d) of the **travelling block.**
5 ___ The **travelling block** is attached to	e) to the top of the **derrick,** and rotates.
6 ___ A **hook** is fixed to the bottom	f) the tower (the top part) of the **oil rig.**
7 ___ The **hook** below the travelling block	g) is attached to a **swivel.**
8 ___ The **top part of the swivel** can't rotate,	h) through a **turntable.**
9 ___ The lower part of the **swivel** is attached	i) and makes it rotate.
10 ___ The **kelly** fits into and goes	j) bottom of the **drill pipe** above the drill bit.
11 ___ The **kelly** is attached below	k) but the **bottom part** can.
12 ___ A **diesel engine** drives the **turntable**	l) the bottom of the **drill pipe.**
13 ___ The **drill bit** is attached to	m) to a 4-sided or 6-sided **kelly.**
14 ___ A **drill collar** is fitted over the	n) the turntable to the **drill pipe.**

2 Describe the stages of a horizontal directional drilling project. Complete the sentences below, using the verbs in the box in the past simple passive.

award choose drill drive enlarge lay miss pull push rotate take

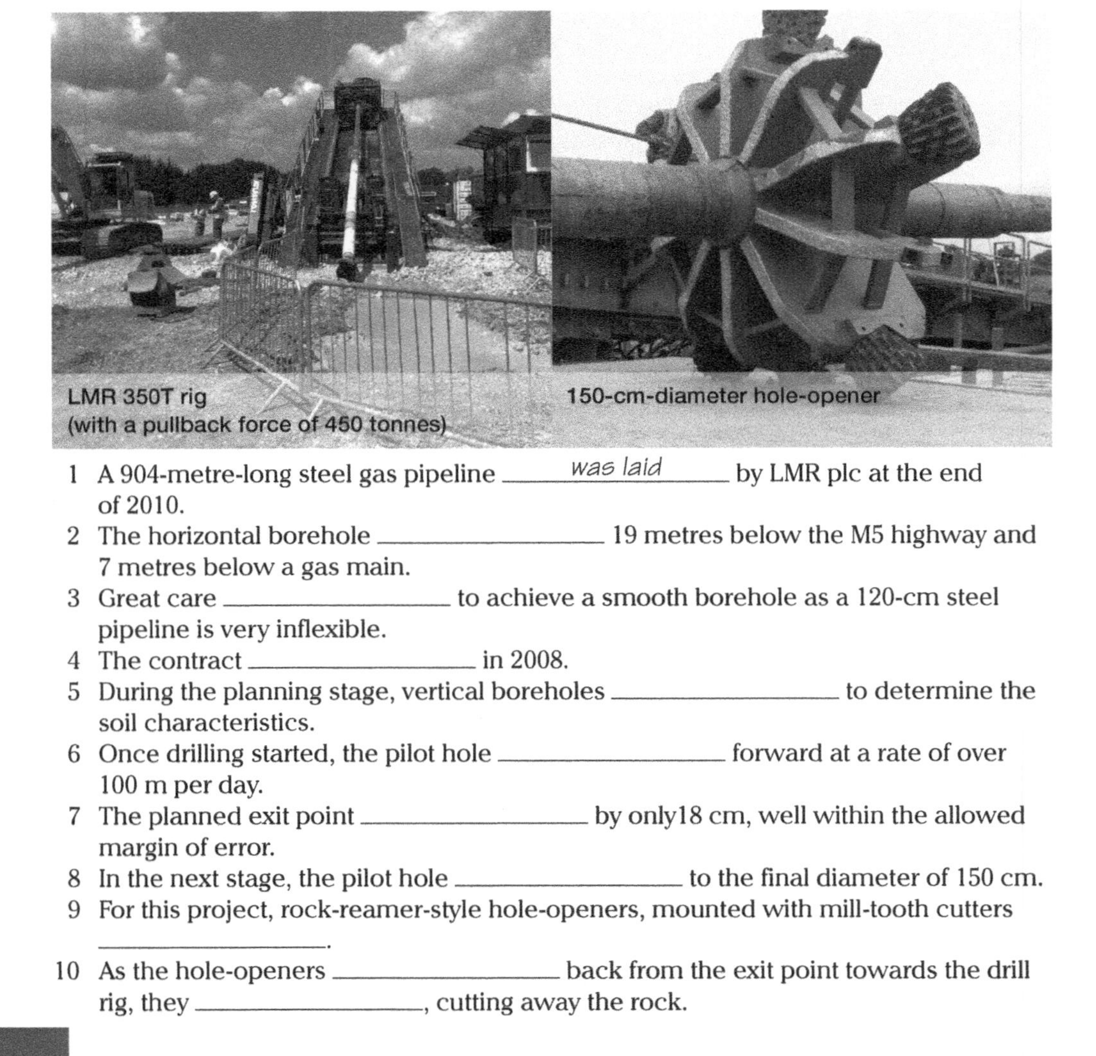
LMR 350T rig (with a pullback force of 450 tonnes)

150-cm-diameter hole-opener

1 A 904-metre-long steel gas pipeline _was laid_ by LMR plc at the end of 2010.
2 The horizontal borehole ________ 19 metres below the M5 highway and 7 metres below a gas main.
3 Great care ________ to achieve a smooth borehole as a 120-cm steel pipeline is very inflexible.
4 The contract ________ in 2008.
5 During the planning stage, vertical boreholes ________ to determine the soil characteristics.
6 Once drilling started, the pilot hole ________ forward at a rate of over 100 m per day.
7 The planned exit point ________ by only18 cm, well within the allowed margin of error.
8 In the next stage, the pilot hole ________ to the final diameter of 150 cm.
9 For this project, rock-reamer-style hole-openers, mounted with mill-tooth cutters ________.
10 As the hole-openers ________ back from the exit point towards the drill rig, they ________, cutting away the rock.

4 Word list

NOUNS (spar)	NOUNS (drilling)	COMPOUND NOUNS	VERBS
barrel	drilling platform	continuous pour	deploy
mooring line	hook	hydraulic cylinder	embed
pipeline	kelly	oil rig	range
pumping station	matrix	recoverable reserves	secure
riser	mud hose		stabilise
seabed	mud pump	reinforced concrete	tow
spar	obstruction		trip in
topside	pilot hole	reinforcing steel	withstand
tree	rebar	rock layer	**ADVERB**
NOUNS (drilling)	rotary table	**ADJECTIVES**	continuously
borehole	slip-forming	complex	
brace	stability	directional	
crown block	swivel	offshore	
derrick	travelling block	sub-sea	
drill bit	terminal	sufficient	
drill collar	turntable		
drill pipe	winch		
drill string			

1 Complete this description of the drilling process, using nouns from the Word list.

The (1) ___*derrick*___ is the tower of the (2) ___*oil rig*___, and stands on the (3) __________. At the top of the derrick is the (4) __________. This rotates when the (5) __________ pulls or releases the cable.

Below the crown block is the (6) __________, which moves up and down and raises or lowers the (7) __________ with the drilling equipment.

The hook is at the bottom of the travelling block and is attached to a (8) __________, the top part of which can't rotate, while the bottom part can.

The lower part of the swivel is attached to a (9) __________, which fits into and goes through a (10) __________. Below this, the kelly is attached to the (11) __________.

At the bottom of the oil well, a heavy (12) __________ fits over the drill pipe just above the drill bit and helps to weigh it down. When the diesel engines are running, the turntable, the kelly and the drill pipe rotate, making the drill bit turn and cut into the rock.

2 22 Write the words in the box on the correct lines. Then listen and check.

continuously cylinder directional drilling hydraulic obstruction pipeline stabilise stability sufficient travelling turntable

1st syllable stressed	2nd syllable stressed
	continuously

D Review

Section 1

1 Change these sentences to reported speech or reported instructions. Use a different reporting verb from the box for each sentence.

tell inform order instruct assure confirm explain promise

Example: *1 She informed him that he had to buy three seats together because of his broken leg.*

1 She said to him, 'Sorry, sir. You must buy three seats together because of your broken leg.'
2 The steward said to the passenger, 'Please don't smoke in the departure lounge.'
3 The helpline said to us, 'Oh, yes, you can carry 10 kg of medical equipment free of charge.'
4 The airline official said to her, 'Don't worry, the plane will not leave without you.'
5 The check-in clerk said to us, 'I'll book you on the next possible flight.'
6 The clerk said to him, 'Folding wheelchairs are carried free of charge.'
7 The steward said to the children, 'Don't sit in the emergency exit row.'
8 The pilot said to the crew, 'Evacuate the aircraft immediately.'

2 Use the words and phrases from the diagram and the box to complete the description of an AC generator.

coil commutator electromagnet north pole south pole

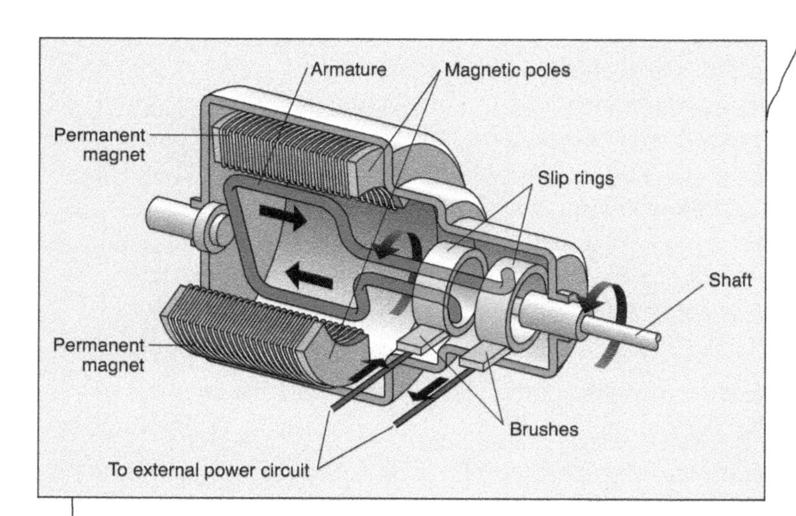

An electrical generator converts mechanical energy into electrical energy. An AC generator, or alternator, is used in a power station to create AC current. A generator consists of a (1) *permanent magnet*, with concave cylindrical (2) ________ which produce a radial magnetic field; and an (3) ________, with a solid steel bar, which is called an (4) ________ and which has a (5) ________ of copper wound around it. The ends of this armature are connected to two (6) ________, which rotate along with the coil. These are made of metal and are insulated from each other. There are two fixed steel (7) ________. One end of each of these is in contact with one of the rotating slip rings, and the other end is connected to the (8) ________.

As the (9) ________ rotates, the armature rotates about its axis and cuts across the force lines of the magnetic field. If the outer terminals of the armature are connected to an external power circuit, an electric current flows through it. During the first half turn, the coil cuts across the field near the magnet's (10) ________ and the lower slip ring becomes positively charged. When the coil cuts near the (11) ________ during the second half turn, the lower slip ring becomes negatively charged. Thus in one rotation of the coil, the current changes its direction twice (alternating current). The faster the coil turns, the greater the frequency (in Hertz) of the current produced by the generator. In a direct current (DC) generator (not illustrated), a split-ring (12) ________ is used to produce a unidirectional current.

Section 2

1 Label items 1–6 in the diagram with the words in the box.

kelly swivel drill string drill collar drill bit casing

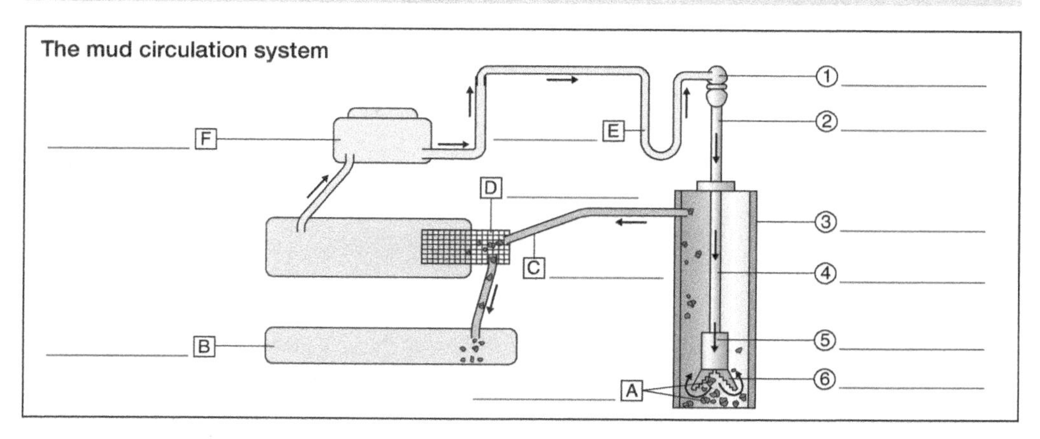

2 Read this text and label items A–F in the diagram in 1, using the words in italics.

An oil well is cleaned out by means of a special liquid called 'drilling mud', which is pumped down the drill string into the well hole. This liquid cools the drill bit, picks up pieces of rock (*rock cuttings*) and carries them up the well hole to the top.
The *mud pump* sucks the drilling mud from large open tanks (called 'mud pits') and pumps it through a *hose*, through the kelly and then down the drill string to the drill bit. There the mud collects the rock cuttings and carries them up through the space between the drill string and the casing. At the top of the well, the mud and cuttings are carried away through the *mud return line*.
Now the mud flows into the *shaker*, which separates the mud from the rock cuttings. The mud goes back into the mud pits and is reused. The rock cuttings go into the *reserve pit*.

3 Write the general words in the box under the illustrations to make technical compound nouns.

arm face field head jacket moulding pad plant plate pulse
reaction shoe sleeve sock

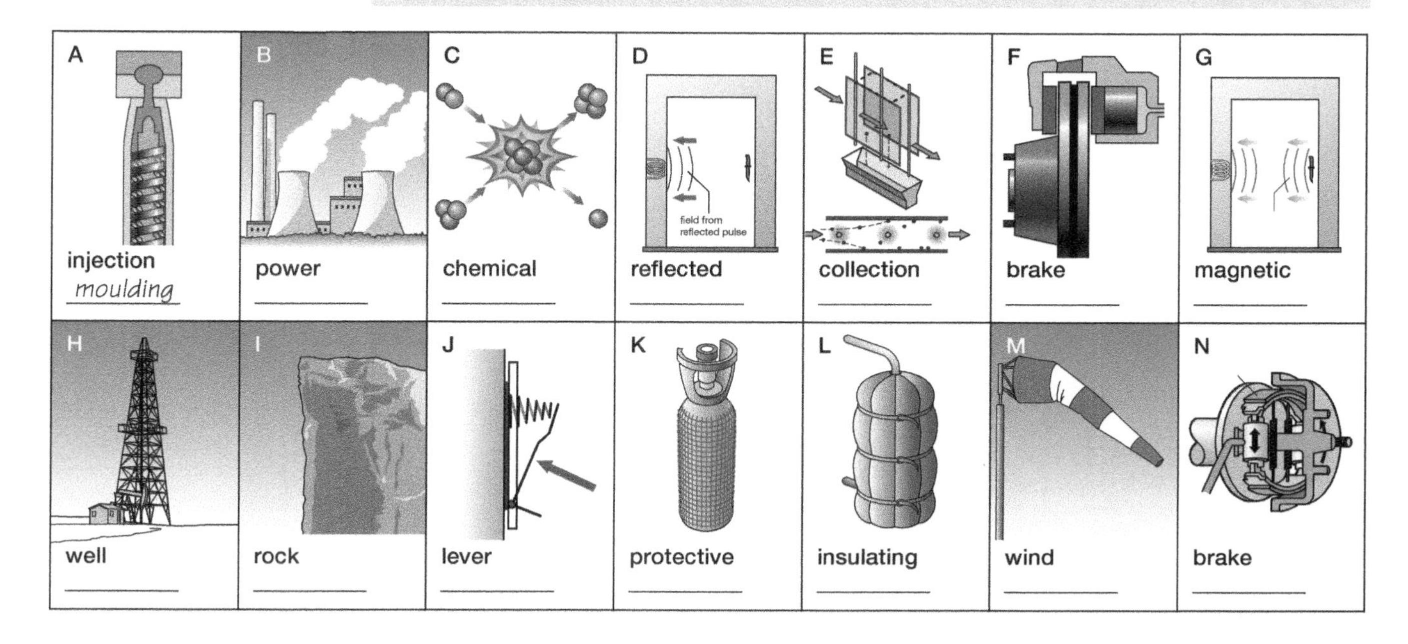

9 Design

1 Inventions

1 Compare the differences between these two sports cars. Modify comparisons 1–4 below in a general way and comparisons 5–8 in a specific way, using the prompts.

	Roadster	GT Sport
1 Acceleration (0–100 kph / secs)	7.9	5.9
2 Fuel economy (kilometres per litre)	15.8	11.6
3 Stability	****	*****
4 Suspension	*****	****
5 Top speed (kph)	177	248
6 Power (brake horsepower)	157	246
7 Storage space (litres)	150	290
8 Price (€)	23,809	34,182

Examples:
1 The GT Sport accelerates a great deal faster than the Roadster.
5 The maximum speed of the GT Sport is 71 kph faster than the Roadster's.

1 GT Sport / accelerate / great deal / fast / Roadster
2 Roadster / lot / economical / GT Sport
3 GT Sport / little / stable / Roadster
4 Roadster's suspension / slightly / good / GT Sport's
5 maximum speed / GT Sport / 71 kph / fast / Roadster's
6 Roadster's engine / two thirds / powerful / GT Sport's
7 GT Sport's storage space / twice / large / Roadster's
8 GT Sport / 50% / expensive / Roadster

2 Write eight more sentences about the cars in 1, using the prompts. Modify your comparisons in a general or specific way as appropriate.

1 acceleration / worse *The Roadster's acceleration is much worse than the GT Sport's.*
2 two thirds / fuel ______
3 unstable ______
4 less comfortable ______
5 two thirds ______
6 one and a half times ______
7 half ______
8 30% ______

3 Write personal answers to these questions about travelling by bus, train or car, using general or specific comparisons.

Example: *1 At weekends, I use the bus much more often. On weekdays, I use the train twice as often as the bus.*

1 Which means of transport do you use more often on weekdays and at weekends?
2 In your experience, how do they compare in terms of comfort, convenience, cost and speed?

2 Buildings

1 23 Listen to the descriptions. Write the description number next to the right picture.

2 24 Listen and write notes in the table about buildings A–C in 1.

	A Burj Khalifa	B Taipei 101	C Shanghai World Financial Centre
1 City	*Dubai*		
2 Country		*Taiwan*	
3 Height			
4 Depth of foundations			*80 m*
5 Storeys			
6 Floor area			
7 Shape			*rectangular opening at top*
8 Use	*hotel,*		
9 Further details		*designed to withstand earthquakes and strong winds*	*highest observation deck*

3 Check the answers for 2. Then, for each of lines 3–6 of the table, write two sentences using superlative forms. Use *easily* or *by far* where appropriate.

3 *Burj Khalifa is by far the tallest building of the three.*
3 ____________________
4 ____________________
4 ____________________
5 ____________________
5 ____________________
6 ____________________
6 ____________________

3 Sites

1 Read the four extracts from emails with descriptions of buildings. Match them with four of the buildings A–H below. Complete the descriptions, using the words in the boxes.

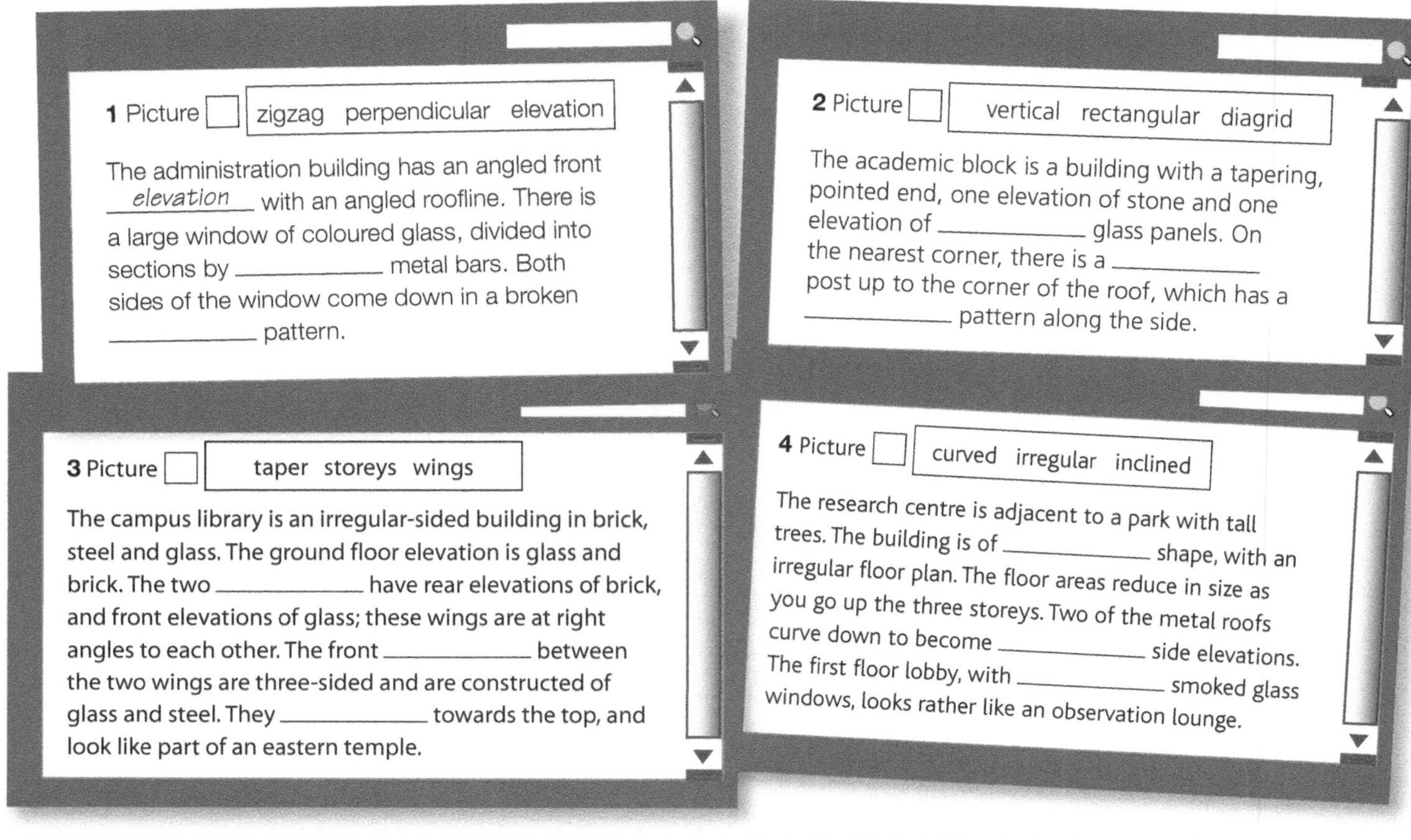

1 Picture ☐ | zigzag perpendicular elevation

The administration building has an angled front *elevation* with an angled roofline. There is a large window of coloured glass, divided into sections by __________ metal bars. Both sides of the window come down in a broken __________ pattern.

2 Picture ☐ | vertical rectangular diagrid

The academic block is a building with a tapering, pointed end, one elevation of stone and one elevation of __________ glass panels. On the nearest corner, there is a __________ post up to the corner of the roof, which has a __________ pattern along the side.

3 Picture ☐ | taper storeys wings

The campus library is an irregular-sided building in brick, steel and glass. The ground floor elevation is glass and brick. The two __________ have rear elevations of brick, and front elevations of glass; these wings are at right angles to each other. The front __________ between the two wings are three-sided and are constructed of glass and steel. They __________ towards the top, and look like part of an eastern temple.

4 Picture ☐ | curved irregular inclined

The research centre is adjacent to a park with tall trees. The building is of __________ shape, with an irregular floor plan. The floor areas reduce in size as you go up the three storeys. Two of the metal roofs curve down to become __________ side elevations. The first floor lobby, with __________ smoked glass windows, looks rather like an observation lounge.

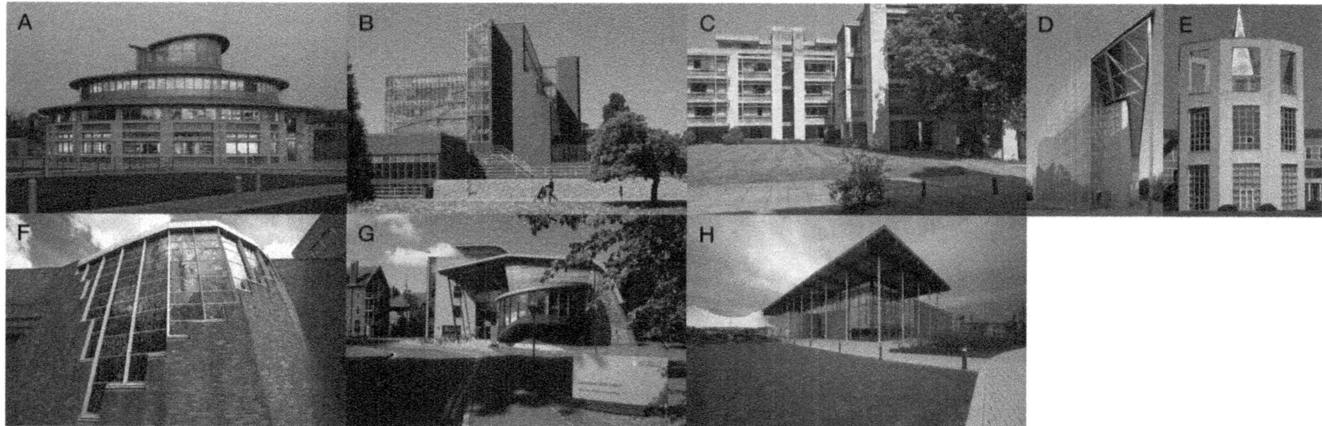

2 Check the answers for 1. Then write four similar descriptions for the remaining buildings, using the prompts. For each one, describe the plan of the building, its elevation and some visible details.

A Department building: semi-circular / circular / elevation / tapering / metal / top / observation lounge

C Student hostel: concrete / vertical / columns / elevation / windows

E Assembly room: three storey / six-sided / rectangular / natural stone / roof terrace / stone framework / conical pyramid / linked / walkways / two levels

H Research centre: rectangular / metal / perpendicular / columns / elevations / glass and brick / left / conical / bicycle racks

4 Word list

NOUNS (products)	NOUNS (buildings)	ADJECTIVES	ADJECTIVES (shapes)
acceleration	aluminium	adjacent	circular
appliance	architect	compact	curved
consumption	beauty	extensive	curvilinear
lumen	concrete	fluorescent	diagonal
stability	elevation	hatched	diagrid
watt	functionality	horizontal	doughnut-shaped
VERBS	innovation	incandescent	
bulge	storey	optional	elliptical
inspire	**ADVERBS**	overall	inclined
taper	approximately	purpose-built	oval
	by far	sail-like	perpendicular
	easily	stable	pointed
	roughly	structural	semi-circular
	virtually	vertical	tapered
			zigzag

1 Label these pictures with adjectives of shape from column 4 of the Word list.

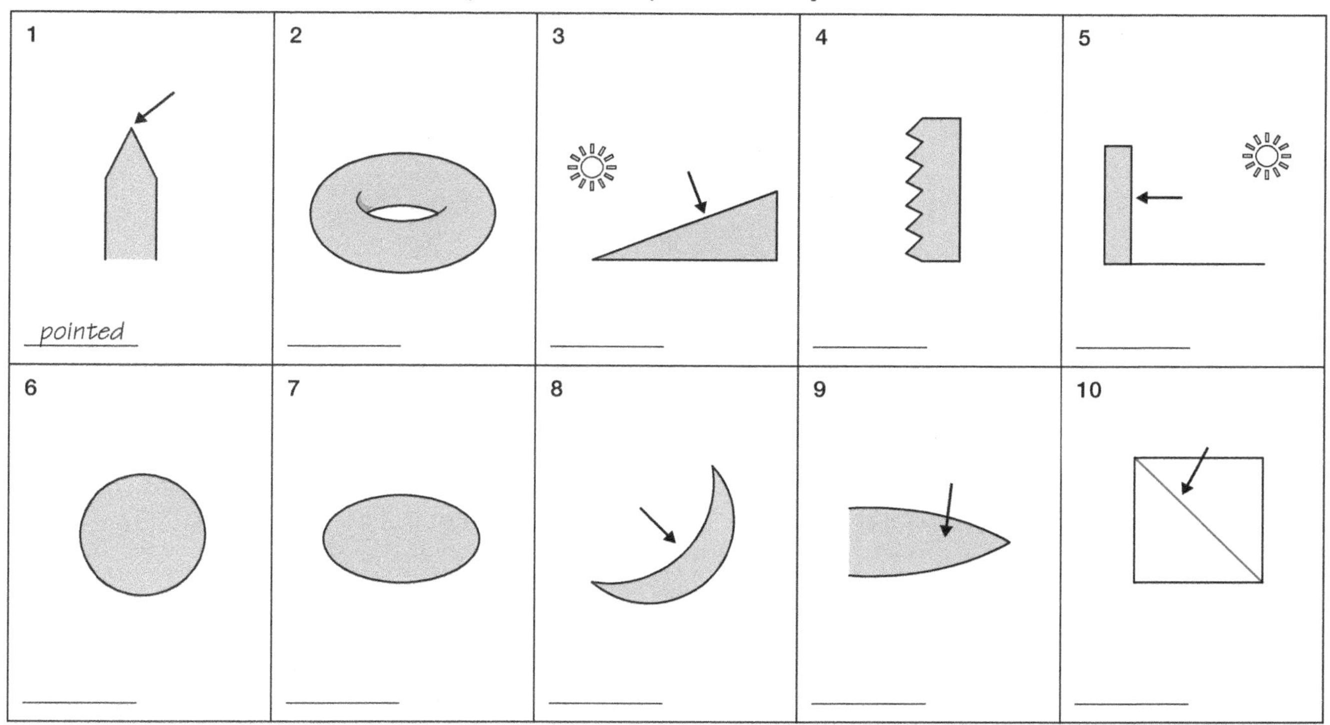

2 Find adjectives in the Word list that are the opposite of the words in italics.

1 a *compulsory* check ____________
2 a *limited* area ____________
3 a *very large* car ____________
4 an *unstable* building ____________
5 a *blunt* instrument ____________

3 25 Listen to and repeat the nouns in column 2 of the Word list and the adjectives in column 4. Underline the syllable with the main stress in each word with more than one syllable.

10 Disasters

1 Speculation

1 Complete this crossword.

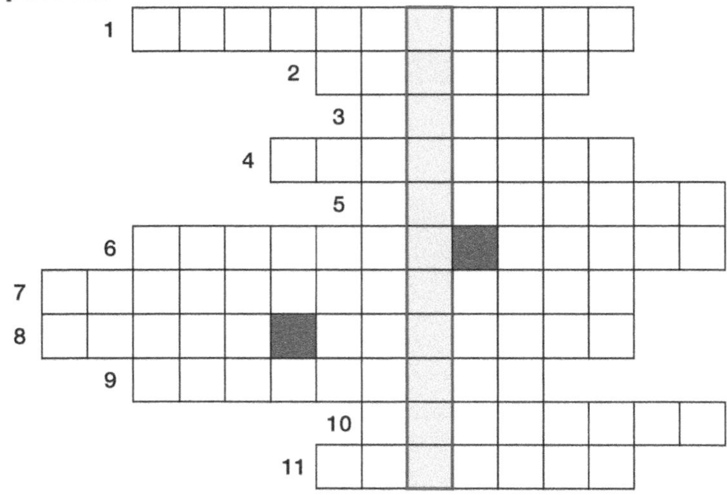

1 pressing together so that the volume is reduced
2 the force of one object striking another
3 removal of surface material by rubbing
4 a break into two or more pieces
5 temporary bending due to compression
6 expansion or contraction because of excessive heat or cold (2 words)
7 discovery and examination of facts about something
8 damage caused by continued loading (2 words)
9 disintegration due to a chemical reaction with water and oxygen
10 sudden falling down (of a structure)
11 a pulling or stretching force

Vertical word: the topic of this section

2 These statements were made after investigations into disasters. Change them into speculations made before the investigations, using the words in brackets. Do not use the words in italics.

Example: *1 The rail crash must have been caused by a broken rail.*

1 The rail crash was *definitely* caused by a broken rail. (must)
2 The rail *possibly* suffered thermal shock due to excessive heat. (may)
3 *Maybe* the aircraft did not have a mechanical failure. (might not)
4 *Perhaps* the flight crew fell asleep due to cabin depressurisation and lack of oxygen. (could)
5 The captain *probably* wanted to save time by steering close to the headland. (might)
6 He *possibly* did not realise that the depth of water in the channel was insufficient. (may)
7 The houses *definitely* didn't collapse in the earthquake because of a design error. (can't)
8 The disintegration of the bridge was *probably* not caused by substandard concrete. (might not)

2 Investigation

1 ▶ 26 Listen to a fictional interview between an accident investigator and a maintenance employee. Write notes or underline the correct alternatives in the table.

Action	Notes	Yes / No
1 New system of rail network checking put in place post-1998		*Yes / No*
2 Speed restrictions put in place post-1998		*Yes / No*
3 Safety valued as a priority over punctuality post-1998		*Yes / No*
4 Date of inspection: rail first identified for repair		
5 Company responsible for replacing rail	*own / other company*	
6 Reason for delay		
7 Length of delay		
8 Previous system for dealing with gauge corner cracking	*rail replacement / preventative grinding*	
9 Countries that use preventative grinding		
10 Number of grinding machines ordered		
11 Total cost	£	

2 ▶ 27 Complete part of the interview in 1, using the words in the box. Then listen again and check your answers.

been should have hadn't wouldn't

A: Another matter, now. Records show that your company identified the rail for repair 21 months before the crash.
B: Yes, we noted the problem.
A: So the faulty rail was discovered 21 months before the crash. But nothing was done?
B Not exactly –
A: Why wasn't the job completed?
B: There was a backlog of essential maintenance work, which was waiting to be done. Also, this job (1) ____________ ____________ ____________ carried out by another contractor, who had agreed to do the work. But they didn't do it.
A: Was *anything* done?
B: Yes, they delivered a replacement rail and left it alongside the track adjacent to the faulty rail.
A: And how long was the replacement rail lying there?
B: I believe it was there for six months.
A: Six months! So if the company that was responsible for the maintenance (2) ____________ delayed for so long, the accident (3) ____________ ____________ happened.
B: That's right. The faulty rail caused the train crash. It (4) ____________ ____________ ____________ repaired, but it wasn't.

3 Reports

1 Complete this abstract of a report into a serious fire at a fuel depot. Put the verbs in brackets into the active or passive, using *must* and *should,* plus the infinitive or perfect infinitive.

This report presents the results of the investigation into the fire at the Buncefield fuel depot. It describes the method of investigation, including CCTV footage, wreckage on site, photos, interviews and examination of pipeline pumping data, both upstream and downstream of the Buncefield depot.

Findings

1. An alarm (1) *should have sounded* (sound) when the liquid in the tank reached its maximum level.
2. The automatic shut-off system (2) ______________ (fail) to stop unleaded petrol being discharged into the tank once it had been filled.
3. Petrol spilled unnoticed from Tank 912 for 40 minutes, generating a cloud of explosive vapour.

Conclusions

The main cause of the fire (3) ______________ (be) the overflowing of around 6,500 barrels of unleaded petrol, after too much fuel was pumped into a storage tank. The investigation team is unable to conclude whether it (4) ______________ (place) the blame either on the design of the alarm system or the operators who were on duty.

Recommendations

1. Back-up sensors that monitor fuel levels and temperatures in the storage tanks (5) ______________ (check) monthly.
2. A new automatic shut-off system (6) ______________ (install) at similar sites.
3. The Health and Safety Authority (7) ______________ (inspect) similar fuel depots immediately to check whether separation distances between fuel tanks are adequate or (8) ______________ (increase).

2 Decide which sections of the fire investigation report would contain the sentences below. Match the sentences to these section headings.

Introduction ☐ ☐	Findings *a* ☐	Background ☐ ☐
Conclusions ☐ ☐	Method ☐ ☐	Recommendations ☐ ☐

a) There must have been a malfunction of the fuel shut-off system, causing unleaded petrol to continue to be discharged into the full tank.
b) In order to carry out a full investigation, the team used a full range of procedures, including interviews with site personnel and examination of CCTV footage.
c) At 06.01 on Sunday, 11th December 2005, the first of a series of explosions started a fire that burned for five days and destroyed almost the entire depot.
d) It is clear that if the second independent alarm linked to warning lights in the control room had functioned, the overflow of fuel would have been detected.
e) During the investigation, it was discovered that automatic shutdown of the pipeline did not take place.
f) The purpose of this document is to present the preliminary results of the investigation into the fire at the Buncefield fuel depot on 11th December 2005.
g) Training courses for control room operatives should be updated.
h) After studying all the evidence, it is not clear whether the source of ignition might have been a spark from an electricity generator or a passing car.
i) An immediate investigation into the causes of the fire was ordered and was formally established by the Health and Safety Authority.
j) Because of the destruction of the control room, it was not possible to study recorded data from site instruments.
k) Tests on all site alarm systems should be carried out weekly.
l) As a result of the fire, fuel supplies to the south of England, including Heathrow airport, were severely disrupted.

4 Word list

NOUNS (general)	NOUNS (structural)	COMPOUND NOUNS	VERBS
abstract	bearing	deck truss	buckle
bomb	buckling	design error	collapse
portion	collapse	gusset plate	compress
procedure	compression	metal fatigue	contract
wreckage	corrosion	support strut	corrode
ADJECTIVES	deck	thermal shock	disintegrate
catastrophic	disintegration		expand
compressive	fracture		fracture
excessive	friction		load
faulty	girder		rust
inadequate	impact		snap
non-destructive	load		spread
significant	node		
temporary	pier		
tensile	rust		
undersized	seal		
	tension		
	walkway		

1 Complete the table with words from the Word list.

NOUNS	VERBS	ADJECTIVES
catastrophe		
compression		
excess	exceed	
destruction	destroy	
buckling		
collapse		
		corrosive
	disintegrate	
	fracture	
		rusty

2 Match the verbs 1–8 with their definitions a–h.

1 d buckle	a) to break suddenly with a sharp noise
2 ___ contract	b) to break up into small pieces
3 ___ corrode	c) to be covered by a reddish-brown substance
4 ___ disintegrate	d) to become bent, e.g. from pressure
5 ___ expand	e) to become bigger, e.g. from heating
6 ___ fracture	f) to be slowly destroyed, e.g. by a chemical
7 ___ rust	g) to break or crack
8 ___ snap	h) to become smaller, e.g. from cooling

3 28 Listen to and repeat the adjectives in column 1 of the Word list. Underline the syllable with the main stress in each adjective.

E Review

Section 1

1 Write six comparative sentences about the two stadium floodlights in columns 1 (RS 90) and 2 (AX 210). Use the information in rows 1–6. Modify the comparisons in a general way, using *far*, *much*, *a great deal*, *a lot*, *slightly* and *a little*.

Example: *1 The AX 210 lamp has a far higher wattage than the RS 90.*

	RS 90	AX 210	KH 240
1 Lamp wattage	1 kW	2 kW	2 kW
2 Lamp lumen output	90,000	210,000	240,000
3 Weight	11.2 kg	18 kg	20 kg
4 Dimensions in mm (length x width)	491 x 276	622 x 520	700 x 420
5 Beam	narrow	medium	wide
6 Control over aiming point	*****	***	*

2 Write six comparative sentences about the stadium floodlights in 1, using the information in rows 1–6. Modify the comparisons in a more specific way, using the prompts.

Example: *1 The wattage of the AX 210 is exactly twice as powerful as the wattage of the RS 90.*

1 wattage / AX 210 / exactly / powerful / RS 90

____

2 lumen output / KH 240 / approximately / 10% / AX 210

____

3 RS 90 / roughly / half / heavy / KH 240

____

4 AX 210 / 100 mm / wide / KH 240

____

5 beam / KH 240 / about / ten times / wide / RS 90

____

6 RS 90 / five times / control / KH 240

____

3 Make six statements about the stadium floodlights, using the information in rows 1–6. Use superlative adjectives and *easily* or *by far*.

Example: *1 The RS 90 has by far the least powerful wattage.*

4 Write the adjectives in the box on the correct line.

circular conical curved curvilinear cylindrical diagonal
diagrid doughnut-shaped elliptical oval perpendicular
pointed semi-circular tapered zigzag

1 **Shapes with curved lines:** *circular,* ______________________________

____

2 **Shapes with straight lines:** *diagonal,* ______________________________

____

3 **Shapes with curved and straight lines:** *conical,* ______________________________

Section 2

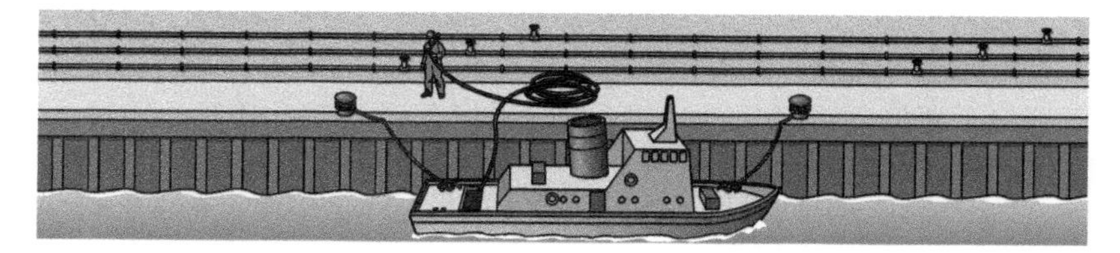

1 Match the sections of the investigation report A–F with these headings.

Introduction	☐	Background	☐	Method	A
Findings	☐	Conclusions	☐	Recommendations	☐

A

The foreman and the surviving members of the maintenance crew were interviewed, as was the stores manager. Time sheets were examined. Photos of the vessel's interior were taken, and the maintenance equipment and tools were examined and photographed.

B

After studying all the evidence, the investigating team has decided that (1) the fire happened because of a build-up of flammable gases below decks. The piping of oxygen below decks had increased the percentage content of oxygen.

As indicated earlier, (2) the accident happened because Worker A did not receive clear instructions and was not properly supervised by the foreman.

C

The purpose of this report is to present the results of the investigation into the fatal fire at Portsmouth dockyard on 30th September 2003, in which three maintenance workers were killed.

D

(a) *Temporary maintenance and repair workers must receive training before starting work.*

(b) *Temporary maintenance and repair workers must be properly supervised.*

(c) *All dockyard gas ring mains* (compressed air, oxygen, acetylene) *must be securely labelled and colour-coded* next to junction points and taps.

(d) *Bulk oxygen should be treated at the supplier's depot before delivery* so that its smell can be detected.

E

At about 11.55 am on Berth C in Portsmouth dockyard, there was a catastrophic flash fire on the vessel *SS Marianna*. As a result, three maintenance workers were fatally injured, and two more workers received serious burns. The vessel did not catch fire following the blaze and was not structurally damaged, although some internal fittings were blackened. An immediate investigation into the cause of the fire was ordered.

F

During the investigation, the following evidence was discovered:

- The foreman told a temporary worker (Worker A) to attach an airline to the dockyard ring main for compressed air, and (3) then went on a meal break, without supervising Worker A.
- Worker A was unable to attach the airline to the ring main, because the manifold was not compatible. He went to the stores and (4) obtained a suitable fitting, connected the airline to the ring main, turned on the supply and returned below deck.
- *Worker B*, who was uninjured, (e) *lit a cigarette* (which was against safety regulations). The cigarette flared up unusually. About five minutes later, he reported feeling unwell, so (5) went above deck to get some fresh air, and thus escaped the fire.
- A few minutes later, there was a major fire below deck, which burnt itself out within a minute.
- Fire officers discovered that the airline had been connected in error to the oxygen ring main, instead of to the compressed air ring main.

2 Rewrite the underlined parts of the report (1–5) as sentences using the third conditional.
Example: *1 The fire would not have happened if there had not been a build-up of flammable gases.*

3 Rewrite the italicised parts of the report (a–e) as sentences using *should (not)* and the perfect infinitive.
Example: *a) Temporary maintenance and repair workers should have received training before starting work.*

11 Materials

1 Equipment

1 Look at the information about different materials and their properties. Write sentences about them in the table below, using the appropriate language.

	burn	bend	stretch	break	absorb impact	resist impact
aramid fibre			a little	no		yes
carbon fibre		yes		no		
polyurethane foam					yes	
thermoplastic polyurethane (TPU)		yes	yes	no		
nylon synthetic fibre			a little	no		
wood	yes	yes	no			
metal		yes		no		
rubber		yes	yes	no		

Language	Sentences
present simple active	1 (aramid fibre) ______ *Aramid fibre resists impact.* 2 (polyurethane foam) ______
can / can't + active	3 (carbon fibre) ______ 4 (rubber) ______
can / can't + passive	5 (aramid fibre) ______ 6 (TPU) ______
active with passive meaning	7 (nylon synthetic fibre) ______ 8 (wood) ______

Plym running shoes

Lightweight, durable, flexible thermoplastic polyurethane: makes the shoe strong and flexible.

Foam-blown polyurethane for extra comfort: tiny air bubbles provide cushioning and absorb shocks.

Mesh added to the upper part of the shoe: allows the foot to breathe.

Increased support and elasticity around the heels and under the soles: allows runners to run long distances without tiring.

Unit price: US$49.50 (per pair)

Package / Delivery free

Delivery max 2 weeks after receipt of order

Offer open 28 days from receipt of proposal

2 Write a letter of proposal to Mr Ali Said, Manager of the Muscat Athletics Club, offering to supply his club with Plym running shoes. Use the information on the left and these prompts.

Thank you / inviting / proposal / supply club / running shoes
Presentation / last week / demonstrated … you / invited / send proposal
As explained, our running shoes / designed …
Shoes combine …
+ mesh / upper / foot / breathe
Shoes give … allowing …
Details / all materials used / attachment / letter
Company proposes / supply / price …
Package and delivery … delivery dates …
Offer / open …
I look forward / order

2 Properties (1)

1 Read sentences 1–5 about things that Kevlar® is used for. Match them to the properties of Kevlar® a–e that make it suitable for these things.

1 _b_ Kevlar® is used for ice hockey masks; it protects the faces of ice hockey players from the flying puck (rubber disc).
2 ___ Fire officers' gloves contain Kevlar®, which protect their hands from cuts and fire.
3 ___ In oil production, Kevlar® is used to reinforce the risers, the pipes that carry the oil from the ocean floor back up to the production platform.
4 ___ Aircraft cabin floors are built with lightweight, honeycomb-core Kevlar® paper, which is fire-resistant and does not easily transmit noise.
5 ___ Snowboard manufacturers use Kevlar® to increase board stability and reduce vibration and weight.

a) It provides lightweight rigidity.
b) It has high impact resistance.
c) It is non-flammable and soundproof.
d) It is heat-resistant and abrasion-resistant.
e) It is flexible and waterproof.

2 Complete the test descriptions using the nouns in the box.

absorbency ductility durability malleability flammability rigidity

1 The _flammability_ test determines the ability of materials to catch fire, to release heat, and to develop smoke under test conditions.
2 In the ___________ test, pieces of different materials were weighed before and after the test. Water was added and the volumes of water taken in by the materials were calculated from the weight increase.
3 In the ___________ test, the plastic laminate is rubbed continuously by an abrasive wheel. The material is examined every 10 hours, and the deterioration of the material is measured. This testing provides an estimate of the plastic product's lifetime span.
4 The ___________ test was carried out by means of a tensile test. A wire was stretched to breaking point, and the percent of elongation (lengthening) was calculated. After the test, the wire retained its changed shape when the load was removed.
5 The ___________ test is similar to a flexibility test. A material is placed repeatedly under increased loads. The test determines whether any bending can be measured, and if so, how much.
6 In a test of ___________, a sheet of the test material is placed on a doming block and is hit repeatedly with a metal punch. The test determines whether the material can be permanently deformed by compression into a new shape without cracking or tearing.

3 Properties (2)

1 29 Listen to two people discussing plans to manufacture a new product. Answer these questions.

1 What product are they thinking of manufacturing?
2 What do they need at the same time as a product specification?
3 When is their next meeting going to be?

2 Listen again and complete the suggestions.

1 ___*I would suggest that*___ we turn our attention to the manufacture of snowboards.
2 ______________ carry out some specific market research into local, national and international markets.
3 ______________ start an R&D programme that could run concurrently?
4 ______________ hiring an R&D specialist with expertise in manufacturing snowboards?
5 ______________ finding someone with the right kind of expertise.
6 ______________ meet again in a week's time to check on progress.

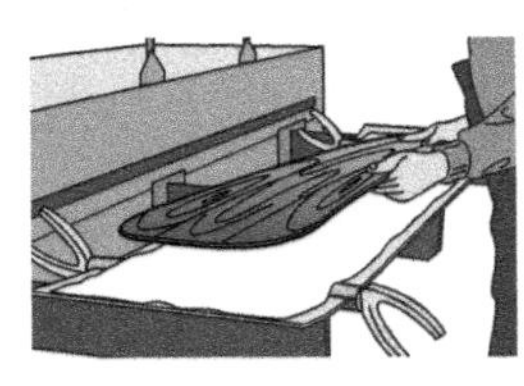

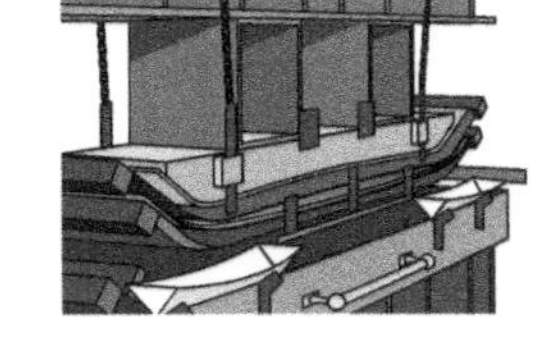

3 30 Listen to a technical presentation about snowboards. Underline the correct alternatives or write notes in the table about the (recommended) options, and the properties.

Component / Feature	Specification options	Properties
1 Wooden core	*softwood* / *hardwood*	a) *long fibres* b) ______ c) ______
2 Laminated fibreglass	*2* / *4* layers	a) ______ b) ______ c) ______
3 Other materials	*carbon fibre* / *Kevlar®* / *aluminium*	a) ______ b) ______
4 Base	polyethylene plastic	
	extruded / *sintered*	*smoother* / *rougher* *faster* / *slower* *porous* / *non-porous* to wax *greater* / *less* surface friction *higher* / *lower* cost *easier* / *harder* to repair

4 Rewrite these sentences using the words in brackets.

Example: *1 Extruded boards cannot absorb wax.*

1 Extruded boards do not have the capacity to absorb wax. (cannot)
2 Fibreglass layers are capable of withstanding torsion and returning to their original shape. (able)
3 A waxed board can reduce drag over all types of snow. (capacity)
4 Sintered boards cannot reach high speeds unless they are waxed. (incapable)
5 The longer board is able to reach faster speeds in downhill races. (capacity)
6 A snowboarder can split the board into two mono-skis. (ability)
7 We have the capability to sell 18,000 units per annum within three years. (capable)
8 However, we are incapable of achieving substantial sales this year. (capability)

4 Word list

ADJECTIVES (properties)	NOUNS (properties)	ADJECTIVES (resistance)	NOUNS (materials)
absorbent	absorbency	impact-absorbent	aramid fibre
ductile	ductility	bulletproof	carbon fibre
durable	durability	childproof	composite
elastic	elasticity	fireproof	graphite
flammable	flammability	ovenproof	Kevlar®
flexible	flexibility	waterproof	nylon
malleable	malleability	corrosion-resistant	polypropylene
non-flammable	non-flammability	impact-resistant	polyurethane foam
plastic	plasticity	shock-resistant	synthetic fibre
rigid	rigidity	heat-tolerant	thermoplastic polyurethane (TPU)
strong in compression	compressive strength	stain-resistant	
		water-resistant	**NOUNS**
strong in tension	tensile strength	**ADJECTIVES**	abrasion
strong in torsion	torsional strength	high-performance	stud
heat resistant	heat resistance	responsive	**VERBS**
strong in shear	shear strength	**ADVERB**	absorb
tolerant	tolerance	incredibly	stretch
lightweight			

1 Complete this table with resistance adjectives from column 3 of the Word list. There may be more than one possible answer.

1 ___*waterproof*___ jacket
2 ______ safety lock in a car
3 ______ laboratory worktop surface
4 ______ jacket for fighting soldiers
5 ______ dish for cooking
6 ______ foam-lined pad
7 ______ boat hull
8 ______ watch for wet weather
9 ______ door, required for hotels
10 ______ camera, for filming on location
11 ______ cooking pot
12 ______ upper of a football boot

2 Match the adjectives 1–8 with their definitions a–h.

1 _*f*_ absorbent	a) can stretch and go back to its original length
2 __ ductile	b) can resist loads without bending
3 __ durable	c) can be rolled or pulled into a longer, thinner shape
4 __ elastic	d) burns easily
5 __ flammable	e) can be permanently formed into a new shape
6 __ flexible	f) can reduce the effect of a sudden impact
7 __ malleable	g) can bend easily without breaking
8 __ rigid	h) stays in good condition for a long time

3 31 Listen to and repeat the highlighted adjectives in column 1 of the Word list, and the highlighted nouns in column 2. Underline the syllable with the main stress in each word.

12 Opportunities

1 Threats

1 Write sentences about the two graphs, using the future perfect of the verb in brackets and *by*, *to* or *at*.

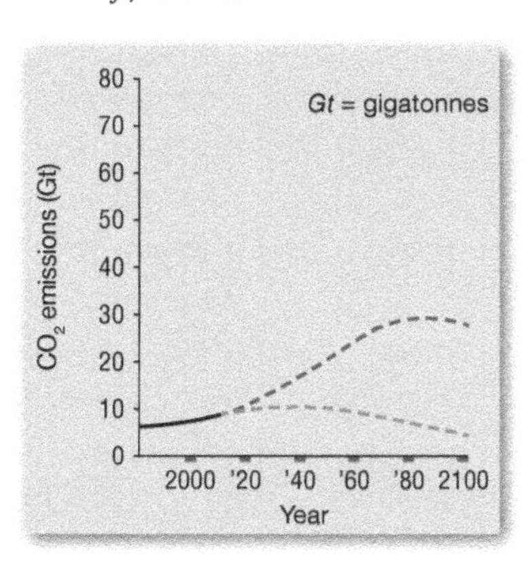

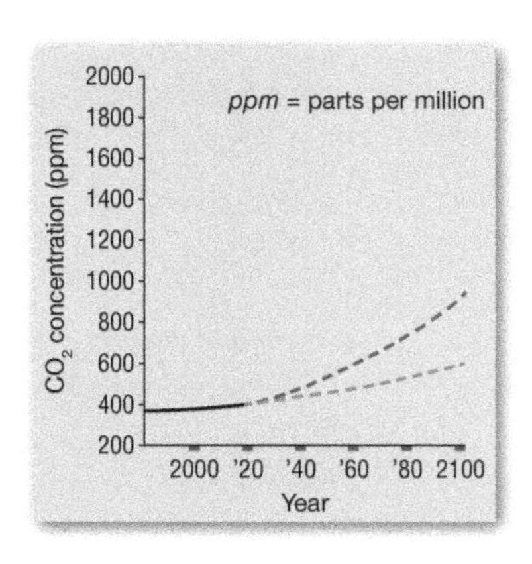

In the worst case scenario

1. By 2040, CO_2 emissions ___*will have risen to*___ 18 gigatonnes per year. (rise)
2. By 2080, CO_2 emissions ___*will have increased by*___ 23 gigatonnes per year. (increase)
3. By 2080, CO_2 concentrations ______________________ 400 parts per million. (grow)
4. By 2100, CO_2 concentrations ______________________ 920 ppm. (climb)

In the best case scenario

5. By 2080, CO_2 emissions ______________________ 6 gigatonnes per annum. (stabilise)
6. By 2100, CO_2 emissions ______________________ 2 gigatonnes p.a. (drop)
7. By 2080, CO_2 concentrations ______________________ 520 parts per million. (increase)
8. By 2100, CO_2 concentrations ______________________ 200 ppm. (rise)

2 Write summaries about what will have happened by the year 2060. Use the prompts, and the future perfect active or passive. A double slash [//] indicates a sentence break.

Example: *1 By 2060, the glaciers south of the Himalayas will have melted. The great rivers coming from the glaciers will have run dry. …*

1. 2060 / glaciers / south of Himalayas / melt // great rivers / coming from glaciers / run dry // all agriculture / river valleys / cease // entire populations / displace
2. 2060 / ice caps / Greenland / Arctic / melt // sea level / rise / by at least 0.4 metres // low-lying areas / flood / some countries / like / Maldives / abandon
3. 2060 / climate change / cause / extreme global warming // many areas / in tropics / as well as / Mediterranean basin / affect / by droughts and fires // many populations / these areas / force / to migrate / southwards and northwards / find / temperate climates
4. 2060 / increased global temperatures / cause / by continued burning / forested areas // by this time / parts / frozen tundra / Siberia / start / to thaw / releasing methane / into / atmosphere // many forests / northeast Asia / destroy / fires / fuelled by methane

2 Innovation

1 Complete the description of a sailboard, using the words in the box.

aerodynamic   apparent wind   direction   drag   friction   inclined
keel   lift   like   polycarbonate   propels   relies   rigid   similar

A sailboard (1) ___relies___ on a flexible nylon and polyester sail, (2) ________ to a sail on a small boat. The flow of wind across the sail (3) ________ the craft forwards. This force, combined with the very low (4) ________, or friction, of high-performance boards, enables the craft to plane across the surface of the water. In high winds, aerodynamic (5) ________ is maximised, while (6) ________ (or drag) is minimised.

A high-performance sailboard is (7) ________ a sailing yacht, in that the wind pushes the sailboard in the (8) ________ of travel. However, in 'displacement sailing' a yacht moves through the water, whereas in aquaplaning, or 'planing', a sailboard skims over the surface at high speeds. In addition, the mast is (9) ________ in the direction of the wind, so the wind lifts the mast and sailboard, thereby reducing drag over the water. This occurs when sailing at high speed at right angles to a strong wind.

A long sailboard is wind-powered, like a sailing boat, but instead of a (10) ________, it has a fixed fin at the rear, and a rigid daggerboard in the centre that can be raised or lowered. The mast is made of light but flexible (11) ________, and the rigid boom is made of aluminium, coated with rubber.

Once a sailboard starts to move, it creates an (12) ________, which can be stronger and faster than the true wind. With the smallest and lightest boards, there is very little surface area in contact with the water and so very little drag. (13) ________ efficiency combined with a (14) ________ board with low drag results in high speeds, with the world speed record currently close to 100 kph.

2 Write sentences about the similarities and differences between these pairs of vehicles, using the characteristics in the table below.

Example: *1 A helicopter doesn't resemble a glider in any way. It uses revolving blades to fly instead of wings. A helicopter is powered, whereas a glider is unpowered.*

1 helicopter / glider
2 submarine / submersible
3 drone / glider
4 space shuttle / helicopter
5 jet ski / hovercraft
6 hovercraft / helicopter

Vehicle	Characteristics	Vehicle	Characteristics
Helicopter	flies at low altitude and can hover; carries passengers and goods; flies by means of revolving blades, not wings	Glider	piloted, unpowered; one- or two-person versions
Space shuttle	piloted; carries out research in space; has wings and lands on a runway	Drone	powered, unmanned small plane, guided by remote control; used for photography and defence purposes
Submarine	manned; propels itself under water; used when submerged for defence purposes	Jet ski	travels over water, propelled by water thrust; can carry one passenger but not goods
Submersible	unmanned device; carries out underwater operations or research tests; can reach extreme ocean depths	Hovercraft	hovers on a cushion of air; travels over land, water and marshy ground; carries passengers, goods and vehicles

3 Priorities

1 [CD 32] Listen to a marketing meeting and complete columns A–D of the table with *yes* or *no* to indicate progress in the car manufacturing industry. Write the name of the fifth system.

System	A R&D	B Prototype testing	C Manufacture in progress	D Infrastructure in place	E Notes
1 solar power	*yes*	*yes*		X	cost coming down at a rate of ______ % p.a.
2 battery exchange (switch station)					countries currently using switch stations: ______
3 battery recharging (domestic plug-in)					units to be produced the year after next: ______ range: ______ km top speed: ______ kph recharging speed: up to ______ % in less than 30 minutes
4 hydrogen fuel cell					X
5 ______	*yes*				increase in models: from ______ to ______ growth in market share: from ______ to ______ %

2 Listen again and complete the notes in column E of the table in 1.

3 [CD 33] Listen and make notes on the second part of the meeting. Write the systems from 1 that the speaker predicts will be best for future large-scale use.

Type of energy zone	Short-term	Medium-term	Long-term
A solar power	*1) solar power*	*1) solar power*	
B hydro-electric power; geothermal power			
c nuclear power			

4 Answer these questions about the systems in 1. Then listen again and check. Which system …

1 is the least expensive to run? *solar power*
2 is the least harmful system for the environment? ______
3 is the cheapest system to buy? ______
4 requires an initial investment in infrastructure? ______
5 is the most convenient system for car drivers? ______
6 requires the greatest investment in infrastructure? ______

4 Word list

NOUNS (climate change)	NOUNS (innovations)	COMPOUND NOUNS	ADJECTIVES
atmosphere	drag	aerodynamic lift	convenient
concentration	friction	apparent wind	exceptional
cyclone	fusion	battery exchange	high-performance
emission	hovercraft	battery recharging	huge
gigatonne	miracle	best-case scenario	massive
glacier	nanotechnology	hydrogen fuel cell	renewable
ice cap	surfboard	solar panel	similar
pattern	traction	wind turbine	spectacular
prediction	**VERBS**	worst-case scenario	staggering
trend	invest		technological
	maximise		tremendous
	minimise		
	rely		
	resemble		
	smash		
	tip		

1 Underline the correct preposition, or *NONE* if no preposition is required.

1 The company invested *in / onto / (NONE)* a new project.
2 The company hopes to maximise *up / to / (NONE)* its profits by reducing staffing levels.
3 The manager relied *in / on / (NONE)* up-to-date marketing reports.
4 The sailing craft resembled *to / at / (NONE)* a triangle of slim girders.
5 The prototype craft smashed *at / into / (NONE)* the cliff face.
6 The sailing craft tipped over *onto / into / (NONE)* its side.

2 Underline the one noun that cannot be combined with each adjective in bold.

1 **exceptional** efficiency weather atmosphere improvement
2 **huge** increase improvement effort prediction
3 **massive** force impact trend explosion
4 **spectacular** pattern building rise result
5 **staggering** speed record exchange achievement
6 **tremendous** force weakness impact earthquake

3 Complete the text using nouns from column 1 in the singular or plural.

Since the mid 18th century, (1) emissions of CO_2 by industrialised nations have increased. Millions of (2) ________ of CO_2 have been discharged. As a result, the (3) ________ of CO_2 in the Earth's (4) ________ has been steadily increasing. (5) ________ of climate change have been closely observed since the mid 1970s on all continents. For example, mountain (6) ________ have been retreating and the extensive (7) ________ of Greenland and Antarctica have reduced in size. In addition, (8) ________ have become more frequent, due to rising ocean temperatures. These (9) ________ are expected to continue into the foreseeable future.

4 34 Listen to and repeat the nouns in column 1 of the Word list, and the adjectives in column 4.

F Review

Section 1

1 Write answers to these questions using the prompts in brackets, not necessarily in the same order.

1. What's the difference between materials of compressive strength and tensile strength?
   (strong / pulled / pressed)
   The first material is strong when it is pressed, and the second is strong when it is pulled.
2. What's the difference between impact-resistant and impact-absorbent materials?
   (soft / rigid / return / shape / not break) ______________________
3. What's the difference between a water-resistant and a waterproof watch?
   (guaranteed / keep out / water / not) ______________________
4. What's the difference between flammable and non-flammable materials?
   (burn / not burn) ______________________
5. What's the difference between materials with torsional strength and materials with tensile strength?
   (strong / stretched / twisted material) ______________________
6. What's the difference between ductile and malleable materials?
   (rolled / pulled / longer, thinner / shape / new shape)
   ____

2 Complete the interview with the words in the box. Use each word once only.

ability able can can't capable capabilities capacity incapable unable

A: Are you (1) *able* to use your new electric bike to commute to work?
B: Of course. These days, I (2) __________ take the car as the streets are too congested. With my new bike, I (3) __________ get through the traffic and arrive at work on time, without feeling tired.
A: What's the new bike like?
B: It's got excellent (4) __________. It has the (5) __________ to get me up hills. It's (6) __________ of a top speed of 50 kph, and has a range of 80 km on a full charge.
A: Can you tell me how you recharge it?
B: The charging time is a bit slow – six to seven hours. Unfortunately, I'm (7) __________ to recharge my batteries at work. However, the bike has enough battery (8) __________ to get me to work and home again on a single charge.
A: Would it be easy to steal?
B: No! It has a steering wheel lock, and an immobiliser, with the same high-tech key as a modern car. So I think a thief would be (9) __________ of going off with it!

Section 2

1 Make predictions for the year 2040, using the future perfect active or passive of the verbs in brackets.

1 My prediction is that photovoltaic cell installations ___will have been built___ (build) across the Sahara from east to west. Huge extra generating capacity ______________ (add). High voltage DC power lines ______________ (install) to allow energy export to Europe, with low line-losses.

2 Biometrics in general and iris recognition technology in particular ______________ (establish) as the new standard security system. People will use it for IT instead of passwords. In the developed world, everybody ______________ (issue) with biometrical documents for travel, banking and personal ID, e.g. passports and driving licences.

3 It's very likely that water export schemes ______________ (become) common, to allow water exports from temperate countries to arid Middle Eastern countries. Glasshouse horticulture ______________ (expand), to allow conservation and recycling of water.

4 I expect that scientists ______________ (develop) new genetically modified (GM) crops that are heat-tolerant and drought-resistant. Those countries that had previously banned GM crops ______________ (change) their policies.

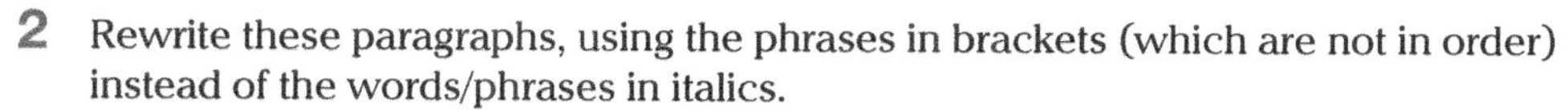

2 Rewrite these paragraphs, using the phrases in brackets (which are not in order) instead of the words/phrases in italics.

Example: *1 … is 25 metres long, instead of the usual 18 metres. Another difference is that it is a double-articulated truck …*

1 The Denby supertruck, which is not allowed on the roads in some countries, is 25 metres long, *unlike normal 18-metre trucks. Also,* it is a double-articulated truck and the rear section can be steered separately.
(Another difference is that) (instead of the usual 18 metres)

2 The Archimedes water screw is *like* a traditional waterwheel. *Just as* water flows over a waterwheel and turns a generator, the flow of water along the Archimedes screw makes it rotate and generate electricity. *In contrast to* a traditional hydro-electric turbine, this screw can turn and generate electricity from a small head of water, for example 1–4 metres in height.
(Unlike) (In the same way as) (has exactly the same function as)

3 An unmanned drone, or remotely piloted aircraft, *resembles* an ordinary aircraft. It is controlled by a pair of operators on the ground, *not by* a pilot. One operator moves the controls to fly the drone *in the same way as* a pilot flies a plane, while the other operator checks the monitor screens and zooms in on the ground below the drone.
(just as) (instead of) (is very similar to)

4 Laser drilling for oil is completely *unlike* normal rotational drilling. A laser installation can be set up quickly and comparatively cheaply, *unlike* a traditional derrick. *Furthermore,* the heat from the laser turns the water in the bedrock into steam, which results in breaking up the rock at the base of the well.
(different from) (Another difference is that) (in contrast to)

Audioscript

Unit 1 Systems

02

[L = Lecturer; S1 = Student 1; S2 = Student 2]

L: My talk today is about FDRs and how they work. As you probably know, FDR stands for Flight Data Recorder, and so it's a device used for recording data during a plane's flight. It also helps investigators to work out what went wrong if a plane crashes.

S1: What happens if a plane crashes into the sea and sinks? How do the investigators find the FDR then?

L: FDRs have an underwater locator beacon. There's a sensor on the side of the beacon. When water touches the sensor, it activates the beacon.

S2: Could you give us some details about the transmission?

L: Yes, the beacon sends out pulses at 37.5 kilohertz.

S2: Can it function at the bottom of the ocean?

L: Yes, it can transmit sound from a depth of 14,000 feet.

S1: Do you have to work fast in order to find the FDR?

L: Well, as soon as the beacon starts to transmit, it will continue for 30 days.

S1: And how frequently is the signal transmitted?

L: The beacon sends out a signal once per second.

S2: What kind of battery is used for the beacon?

L: It's powered by a battery with a shelf life of six years.

S1: What happens after you find the FDR?

L: After an air accident, the flight recorder is transported to the computer lab. There, the data can be analysed. If an accident happens at sea and the flight recorder was in the water, the FDR must be transported in a container of water to keep it cool. Any other questions?

S1: I can't see the beacon very clearly from here. What shape is it exactly?

L: It's cylindrical. The beacon also serves as a handle.

S2: I heard that Flight Data Recorders are sometimes called 'black boxes'. Is that right?

L: That's correct, yes.

S2: Then can you explain why they're called 'black boxes' if they're bright orange?

L: Good question. We're not sure why they're called 'black boxes' but we do know that the early FDRs were in fact black, because they used a film-based technology. So the insides of the boxes had to be black to stop any light affecting the film. Nowadays, they're painted bright orange, so that they're easier to find.

03

Hi, everyone. I'm going to demonstrate this self-inflating life raft. This one's big enough for eight people, though there are bigger ones for 12 people, or small ones that take only four.

So, let's take a closer look at this life raft. On top, you've got a canopy to keep out the waves and the rain. This canopy inflates automatically and it's insulated to keep the occupants warm. The floor of the raft is insulated, too. The canopy has lights on top, which are water-activated. It also has a device for collecting rainwater.

Now let's have a look at the automatic inflation system. First, the two buoyancy chambers are inflated, that's the two circular tyres that form the outer walls of the raft. This forces open the vinyl carrying bag. Then the canopy over the top is inflated. Then you can get in. The whole process of inflation takes 12 seconds, OK? That's just 12 seconds.

Next point, stability. There's a big danger with standard life rafts that the wind will catch the life raft and capsize it, or flip it over. But on this model, there are two big stabilisation chambers under the floor of the life raft. Both chambers quickly fill with water from the moment of inflation. They fill in two stages. First, the upper chamber fills through vertical portholes in the chamber walls. At the same time, the lower chamber starts to fill. Water gets in through the bottom here, through a one-way valve which lets water in, but not out.

Once these stabilisation chambers are full, you've got a stable life raft. You could find yourself in a hurricane, with ten-metre-high waves and winds up to 300 kilometres per hour. Even in a hurricane like that, this life raft is so stable that it's unlikely to capsize. And if a giant wave comes along and flips the life raft over, the weight of the water in the stabilisation chambers will flip the life raft back up again. In other words, it rights itself at once.

Unit 2 Processes

04

Good morning, everyone. This is the last in our series of short talks about the plastics industry. Last week we looked at the process of extrusion blow moulding. Today, I'm going to explain how vacuum moulding works. This is also called vacuum forming, or vacu-forming.

Vacuum forming is a process of making a shape out of a thermoplastic sheet. A thermoplastic sheet is a material that becomes soft when you heat it, and later on it becomes hard after you cool it. In vacuum forming, as in other processes of plastic moulding, heating and cooling are important parts of the process.

As its name suggests, with vacuum forming, we have to create a vacuum in the mould in order to produce the moulded object. As you remember, a vacuum occurs when we suck the air out of a container. So sucking air out, or suction, is an important part of the process, and we use a simple air pump for this purpose.

Now, I've given you a printed handout with some diagrams. As you can see in Figure 2 on your handout, there are some air-exhaust holes at the bottom of the mould. These are for sucking the air out.

But let's start with the first stage, which is heating. As Figure 1 on your handout illustrates, the heaters are situated above the machine and heat the thermoplastic sheet below them.

The next stage is positioning. As can be seen in Figure 2, the thermoplastic sheet is moved to a position directly above the vacuum mould.

The third stage is illustrated in Figure 3. Here, the thermoplastic sheet is stretched over the top part of the mould. Here you can see that the plastic material is stretching and becoming thinner. This is the stretching stage.

The fourth and final stage is shown in Figure 4. Here the pump on the right-hand side of the machine is sucking the air out of the mould, and this suction is creating a vacuum. Finally in this vacuum forming stage, as Figure 4 clearly shows, the plastic sheet is sucked down over the mould.

05

close, closure
compress, compression
eject, ejection
expand, expansion
extrude, extrusion
heat, heater
inflate, inflation
inject, injection
mould, mould
rotate, rotation
transfer [verb], transfer [noun]

Unit 3 Events

06

My talk today is about ejecting from an aircraft. This sounds dangerous, but in fact it's very simple. You only have to pull one handle in order to eject. The ejection system does the rest. You can eject at airspeeds between zero and 1,400 kilometres per hour, and at altitudes between zero and 82,000 feet. You only eject from your aircraft if it's going to crash, so first you must try to save your aircraft. If you know that you cannot control your aircraft, pull the ejection handle.

The ejection system operates in four stages. First, the canopy over the pilot is jettisoned. It flies away from the cockpit to the rear. At the same time, the aircraft services are disconnected, so radio contact is lost. However, the emergency oxygen supply for the pilot comes into operation, so it's still possible to breathe. Next, the ejector rocket fires. This ejects the pilot into the air, but the pilot is still restrained in his seat. So now we have the pilot, still strapped into his seat, in the air and starting to fall.

At this point, stage two begins, and the drogue is deployed. Now a drogue is rather like a parachute. Its purpose is to stabilise and slow the fall of the pilot, who is still in his seat. This continues down to a height of 8,000 feet, when the drogue is released.

We now come to the third stage of the ejection procedure, when several things happen in quick succession one after the other. As the drogue falls away, an explosive charge separates the main parachute container. Also, the main harness that attaches the pilot to his seat is released. Small sticker clips hold the pilot in his seat for a few seconds longer. Immediately after this, the main parachute is deployed. This slows the fall of the pilot and lifts him out of his seat, pulling away the small sticker clips that kept him in his seat up until now. The pilot's seat, which is heavier than the pilot, falls to the ground separately.
Now we come to stage four, the landing. The pilot is falling towards the ground at the rate of a normal parachute descent. At this point, his personal locator beacon is activated automatically and starts to send out a radio signal.

abort engine
attitude-control engine
crew capsule
ejection system
jet fighter
jettison engine
launch abort system
oil rig

Unit 4 Careers

[I = Interviewer; L = Laura]

I: Thank you for coming in for an interview, Ms Gallini. I've read through your CV and I'd just like to check a few details. You've applied for the post of Technician, but we currently have two technician posts vacant. Which job are you interested in?
L: I'm particularly interested in the job of Technician for New Product Development.
I: OK. I see that you're currently working at Horton Engineering as a Junior Technician.
L: Actually, I'm now a Senior Technician, since last month.
I: I see. Good! And how long have you worked there?
L: I joined in 2008, so I've been there since then, until now. I'm sorry, it says 2005 on my CV, but the correct date is 2008.
I: And what are your responsibilities in your current job?
L: Most of the time I work on quality control and product testing. But I'm currently helping develop a new product. That involves building a prototype. It's a company secret, so I can't say much about the project.
I: Fair enough! Now, exactly what kind of business is your company in?
L: We're involved in robotics, and medical engineering as well. So, more or less the same field as yourselves.
I: Mmm. I see you started your career before you went to university. How long did you work at Farley Marine?
L: I worked there for two years, from 2003 until 2005. By the way, they've changed their name. They were taken over a couple of years ago, and now their name is BAMC plc.
I: Right. What was your job description there? And what did you do every day?
L: I was an Apprentice Engineer. Most of the time I did machining and finishing. But at the end I was doing some quality control work, too.
I: Excellent. And then you went on to Albany College of Engineering. How long did your course last?
L: It was a three-year course. I went there in 2005 …
I: And you got your qualification at the end of it, I see, in 2008.
L: That's right. I was awarded a Bachelor of Applied Sciences. The course involved all the sciences, but I specialised in Physics and Mechanical Engineering.
I: And why do you want to leave Horton Engineering?
L: It's a small company, so I'd like to move to a bigger company that has a bigger R&D department and a greater product range. I'd really like to try my hand …

09

1 Which job are you interested in?
2 How long have you worked there?
3 What are your responsibilities in your current job?
4 Exactly what kind of business is your company in?
5 How long did you work at Farley Marine?
6 What was your job description there?
7 How long did your course last?
8 Why do you want to leave Horton Engineering?

First syllable stressed:
benefit
competence
previous

Second syllable stressed:
ambition
apprentice
apprenticeship
certificate
diploma
technician

Third syllable stressed:
engineering
institution
interpersonal

Unit 5 Safety

[C = Customer; H = Helpline]

1

C: Hello. I wonder if you can help me.
H: I'll try!
C: I've just bought a used car – it's a Cordoba 1.8 and it doesn't have an owner's manual. The problem is, I've got a yellow warning light on the dashboard. It's shaped like a warning triangle in a circular arrow.
H: I think I know what that is. It's the ABS.
C: ABS?
H: Yes, the Anti-lock Braking System. How is the car handling? For example, do you need to take any corrective action when you brake?
C: Yes. When I brake heavily, I have to counter-steer to the left.
H: I see. That's the problem. You need to take it in to a dealer.
C: Is it safe to drive?
H: Yes, it is. But you should have it checked out as soon as possible.

2

C: Hello, I've got a problem with my car and I don't know what the warning light on my dashboard means.
H: Where exactly is the warning light and what colour is it?
C: It's in the middle of the warning lights panel and it's red.
H: What shape is it?
C: Well, I'm looking at it now. It's square-shaped, with a bit sticking out.
H: Aah! That indicates that one of your doors isn't closed.
C: I've checked those already.
H: What about the rear door? Have you checked that? If any of the side doors or the rear door isn't securely closed, the warning light will come on.

3

C: Hello. Can you help me with a warning light on my dashboard?
H: Of course. Where is it and what does it look like?
C: It's red and it's in the shape of an oil can.
H: Is it on all the time?
C: No, I've been monitoring it all day. It just comes on from time to time.
H: Right. That's the warning light for oil pressure.
C: Should I change the oil?
H: No, that's not necessary. But you must top up the car with engine oil. Don't drive anywhere until you've done that.

4

C: Hello. Can you help me? I've got a problem with the car that I bought from you last month.
H: Which model?
C: It's a Bari 220S.
H: And what exactly is the problem?
C: I'm getting vibration in the steering wheel. That's when I drive at more than 80 kilometres per hour.
H: Any warning signs on the dashboard?
C: Yes. It's orange and shaped like a tyre.
H: I see. Do you see the LED display on your dashboard? Scroll down to 'tyre sensors' and click.
C: Aah!
H: What's the display now?
C: It says: 'Front nearside tyre, incorrect pressure.'
H: Scroll down a bit.
C: Aah! Now it says: 'Above maximum pressure: deflate.'
H: So that's the problem. The tyre is over-inflated, so you need to let some air out, using a reliable tyre pressure gauge.

5

C: Hello. I may have a problem with my car. I've just bought a used car and a warning light has just lit up on the dashboard. The problem is, there isn't a manual in the car.

H: What colour is the warning light and what does it look like?

C: It's red and it's shaped like a petrol pump.

H: Is it a visual or an audible signal, or both?

C: There's a loud audible warning when I turn on the ignition.

H: What's the fuel gauge showing?

C: It's very low. The needle is pointing into the red section.

H: Then you have less than eight litres in your petrol tank. You need to fill up at once.

1

H: How is the car handling? For example, do you need to take any corrective action when you brake?

C: Yes. When I brake heavily, I have to counter-steer to the left.

2

H: Where exactly is the warning light and what colour is it?

C: It's in the middle of the warning lights panel and it's red.

3

H: Is it on all the time?

C: No, I've been monitoring it all day. It just comes on from time to time.

4

H: And what exactly is the problem?

C: I'm getting vibration in the steering wheel.

5

H: Is it a visual or an audible signal, or both?

C: There's a loud audible warning when I turn on the ignition.

13

A: I sent you a proposal about a new in-car warning system that would warn drivers when they get too close to the vehicle in front. For example, this could be a tactile or audible warning.

B: In other words, you're suggesting a beep or something that makes the steering wheel vibrate.

C: I agree with that. I think we're on the right lines. By the way, talking of lines, I was late today because of a pile-up in the fog. I think a lorry had gone off the road and …

A: Yes, yes. Anyway, you're here now, so let's get back to the subject. Alternatively, we could have a visual warning sign, for instance one that was flashed onto the inside of the driver's windscreen.

B: I don't think that's a good idea. There are too many flashing lights already when you're driving.

A: Well, what about a system based on different sensors, to monitor things like the distance from the vehicle in front, your speed, the road conditions and the external temperature? Then we're combining different factors.

C: That sounds good. So, an audible or tactile warning system related to different factors.

Unit 6 Planning

14

[I = Interviewer; E = Expert]

I: Do you think that carbon emissions will push up global temperatures?

E: We've already had an increase in global temperatures of zero point five degrees Celsius. In future, if no action is taken, we can expect a further rise in global temperatures of between two and three degrees Celsius by 2060.

I: Is that certain?

E: No, but it's a 75% possibility.

I: Is this the worst that could happen in the future?

E: By no means. There's a 50% chance that average global temperatures could rise by five degrees Celsius, also by 2060.

I: So how could we avoid this?

E: Well, we need to stabilise our carbon emissions by 2030. After that, we would need to reduce emissions by between 1% and 3%.

I: I see. What are your views on switching energy sources?

E: In our opinion, it's essential to promote all the technologies for cleaner energy and cleaner transport. We should aim to have non-carbon-based fuels accounting for 60% of energy output by 2050.

I: So will that reduce carbon emissions in the long term?

E: Yes, it will. We need to set two new European targets: to reduce carbon emissions by 30% by 2020, and by 60% by 2050.

I: What was that last figure?

E: Our carbon emissions in 2050 need to be 60% lower than they are now.

I: Thank you for clarifying those parts of your report.

15

1 Do you think that carbon emissions will push up global temperatures?
2 Is this the worst that could happen in the future?
3 So how could we avoid this?
4 What are your views on switching energy sources?
5 So will that reduce carbon emissions in the long term?

16

Verbs	**Compound nouns**
condense	bio fuel
consume	carbon capture and storage
convert	clean coal technology
desulphurise	collection plate
emit	flue gas
gasify	iron filing
humidify	power plant
ionise	
liquefy	
pressurise	
pulverise	
purify	
recover	
solidify	
sulphurise	
switch	

Unit 7 Reports

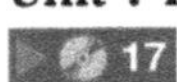

[M = Manager; B = Bob]

M: Hi, Bob. Welcome back. So, how are you getting on?

B: I've made some progress since we last spoke.

M: So, what systems have you come up with?

B: I've looked into four of the identification technologies that you suggested last month.

M: And?

B: As we thought, we need to look at biometrics in order to obtain the fastest and most secure system.

M: And what types of biometrics did you look at?

B: Well, biometrics covers the physical measurement of a person – their voice, their fingers, the shape of their face, or the patterns of their irises. And the most promising technology is iris scanning.

M: Really? What about that other system that we spoke about – capacitive scanning? Have you looked into that?

B: Yes, I'll come to that in a moment.

M: Remind me why biometrics is going to be the best solution.

B: Well, take the human iris. For a start, every human iris is individual. The chances of two different irises being the same are one in ten to the power of 180.

M: I follow.

B: And the iris structure doesn't change during a person's lifetime. Basically, you will have the same iris structure when you are old that you had as a one-year-old baby. Unless of course you have to have eye surgery. That might change the iris structure.

M: So, iris scanning, then.

B: In fact, the basic iris recognition technology is known as 'image capturing', not scanning.

M: What's the difference?

B: Image capturing is video; scanning is, well, scanning.

M: Yes, I see what you mean.

B: With image capturing, there's no physical contact, unlike fingerprint scanning or capacitive scanning. So no germs or bacteria are transferred.

M: Go on, tell me more.

B: It's accurate. And it's much more reliable than voice recognition. And we agreed that accuracy is our first priority.

M: Agreed. Any other good points you discovered?

B: It's flexible. If you want, you can combine iris scanning with other systems, like PIN numbers or swipe cards. Or you can use iris scanning as a stand-alone system.

M: So how does the system operate?

B: In stage 1, the registration stage, you capture an image of the person's iris. Next, you upload all the details of the images plus the person's ID onto the computer. Then, for identification, when the person gets to the access gate, they face the iris scanner, and the computer matches the image of their iris with the details on the database.

M: And how long does all that take?

B: Less than two seconds.

M: I'm impressed! Anything else?

B: Yes, scalability. Once you have the system up and running, you can add thousands of people to the database. It has no effect on accuracy or access time at the gate.

M: So you're saying the system could handle thousands of people.

B: You can have a multi-million person database! This system is going to be huge, I can tell you!

M: Good. Oh! I must go. I'll catch up with you after the weekend.

18

1 How are you getting on?
2 What systems have you come up with?
3 What about that other system that we spoke about?
4 Have you looked into that?
5 Yes, I see what you mean.
6 Go on, tell me more.
7 So how does the system operate?
8 I'll catch up with you after the weekend.

amplifier
capacitance
capacitor
conductor
dielectric
magnetism
microchip
resistor
sensor
terminal

Unit 8 Projects

Hello and welcome to the new Visitors Centre at the Three Gorges Project. The Centre was opened in 2009.

Many visitors ask why the TGP was constructed, so we hope that this short talk will give you some general information. In the past, the Yangtze River used to flood every year, particularly in the flood months between May and September. These floods caused thousands of deaths every year, as well as destroying houses, businesses and farmland. So, this was one reason for constructing the TGP, which has eliminated flooding and loss of life. Since the dam was completed, this has brought a further advantage, since it's now possible to regulate the flow of water through the three gorges, and to even out the water flow during the 12 months of the year. At the earliest planning stage, an important aim of the project was the generating of electricity.

Two power plants were constructed, one on each side of the river, with a combined capacity of 18,200 MW, the largest in the world.

The area covered by the present reservoir once included two cities and 116 towns. About 1.1 million people had to be resettled as the waters of the reservoir began to rise. Many thousands of the displaced persons chose to emigrate to other cities in search of employment. Those that chose to remain in the gorges have all been re-housed, and many new towns have been built for this purpose. These new settlements have been well laid out by providing grassy open spaces around the five-storey housing blocks. The people who have moved there appreciate the quality of their new housing, which is superior to the simple houses that they had before.

Finally, because this huge new dam lies close to an earthquake zone, many visitors ask questions about the safety of the TGP. In fact, the dam has been constructed to a high specification in order to withstand earthquake shocks. Thank you for listening. To hear some statistics about the Three Gorges Project, press '7'.

Here are some statistics about the Three Gorges Project.

The scale of the Three Gorges Dam is enormous, and huge quantities of rock and concrete had to be moved during the construction phase. 103 million cubic metres of earth and rock were excavated and moved. 32 million cubic metres of earth and rock were moved for filling. A total of 28 million tonnes of concrete was mixed, transported and poured. Massive quantities of metalwork were used on the project, including 462,300 tonnes of rebar. There was also a large requirement for general metalwork, for things like gates, beams, and so on. This totalled 256,500 tonnes. The total volume of the dam is estimated at 12 million cubic metres. It's surprising, not that the TGP took so long to construct, but that such a huge project was completed in such a short period, just 17 years. Work on this dam, the largest in the world, started in 1993 and was completed on schedule in 2009. During its construction, about 20,000 people were involved in building the dam.

The dimensions of this dam are as follows: it has a maximum height of 181 metres and is 2,309 metres long; in other words, it is over two kilometres in length.

To build such a huge dam, it was first necessary to construct nine concrete batching plants, for the mixing of the concrete. These had a very large production capacity, and when all nine plants were working, they had the capacity to produce 2,400 cubic metres of concrete per hour. In fact, in one month, these nine batching plants broke a record: they produced 553,000 cubic metres of concrete, that's over half a million cubic metres, in one month! Finally, behind the massive Three Gorges dam lies the reservoir, which covers an area of 1,084 square kilometres. The reservoir is 660 kilometres long. To return to the main menu, press '0'.

First syllable stressed:	Second syllable stressed:
cylinder	continuously
drilling	directional
pipeline	hydraulic
stabilise	obstruction
travelling	stability
turntable	sufficient

Unit 9 Design

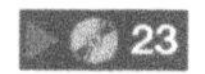

1

The base of the building is square, but for three quarters of its height it tapers to its recognisable rectangular top, where an opening reduces the stresses of wind pressure. Its shape is often compared to a giant bottle-opener. The hotel on the 79th to 93rd floors is the highest in the world.

2

When this building was completed in 2004, it was the highest in the world. The towers remain the tallest twin buildings in the world. The two spines at the top of the building, which are not antennas, form part of the structure. The towers are linked by a two-storey skybridge linking the 41st and 42nd floors.

3

For a time, this was the tallest building in the world. It has an unusual shape: eight tapering sections rest upon a huge base or pedestal, which also tapers upwards from the ground floor, which is in the shape of a square.

4

This building is likely to remain the tallest building in the world for many years. The Y-shaped base gives the structure great stability. The lower part of the building provides 160 floors for human occupation. The top 46 levels of the thin, pointed spire are for maintenance and services only.

Picture A

This was called Burj Dubai – 'Dubai' because of its location here in Dubai in the United Arab Emirates – but they changed the name when it opened to Burj Khalifa. It's the tallest building in the world, at 828 metres, with effectively 206 floors, but the top 46 are only for maintenance. The foundations have to be deep – they go down to 50 metres. The floor area extends to 464,000 square metres.

You probably noticed the shape of the building as you came in. It's quite distinctive and elegant. In fact, it's Y-shaped, which gives the building greater structural stability.

It's a building for mixed use: there's a large hotel, and there are apartments and restaurants and an outdoor observation deck. Do you have any questions? ...

I can give you some more details, if you like. For a start, there are 25,000 windows to clean.

Picture B

Do you see that very tall building in the distance? That building is the tallest building in the city of Taipei and in the whole of Taiwan. It's called Taipei 101. For a time, it was the highest building in the world but it isn't the highest any more. It's 509 metres high, and the foundations go down 80 metres under the ground. That's because it's built in an earthquake zone. So it has to be very stable, both to withstand the earthquakes and also the very high winds that you get in this area.

It has 101 storeys, plus 5 storeys under ground. And the floor area covers 412,000 square metres.

You can see it has a very recognisable shape: there are eight tapered sections of eight floors each. They rest on a tapered base, which is square. Finally, Taipei 101 is a mixed-use building, consisting mainly of offices and conference centres, with some retail shops and restaurants.

Picture C

On your right is the Shanghai World Financial Centre. Althought it isn't the highest building in the world, it's the highest in Shanghai and the highest in China – 494 metres high, and the foundations go down 79 metres under ground. The original design was for a tower of 94 storeys, but the plans were changed and this has 101 storeys, with a floor area of 381,600 square metres. It has an interesting shape, rather like a bottle-opener. It's square at the base, but tapers to a rectangular top. The hole at the top has a very practical purpose and reduces the wind pressure caused by high winds. It's a mixed-use building. There's a hotel near the top of the skyscraper, and there are offices, conference centres and shopping malls on the ground floors. Above the opening at the top is a public observation deck, on the 100th floor, which is the highest in the world.

Nouns (buildings)	**Adjectives (shapes)**
aluminium	circular
architect	curved
beauty	curvilinear
concrete	diagonal
elevation	diagrid
functionality	doughnut-shaped
innovation	elliptical
storey	inclined
	oval
	perpendicular
	pointed
	semi-circular
	tapered
	zigzag

Unit 10 Disasters

[I = Interviewer; E = Employee]

I: ... so it looks as if the crash involved a broken rail. After the derailment in Scotland in 1998, which also involved a broken rail, was the rest of the rail network in the UK inspected?
E: The system of routine checking and maintenance was continued.
I: But was a new schedule of checking the whole network for gauge corner cracking put in place?
E: Not as far as I know.
I: Does that mean 'no'?
E: Yes, it does.
I: I also understand that, after the crash, speed restrictions should have been ordered while the whole network was checked for cracked rails. Why wasn't this done?
E: Speed restrictions would have resulted in delays, and the company didn't want to be responsible for delayed trains.
I: So was punctuality given a greater priority over safety?
E: I guess you could say that.
I: Another matter, now. Records show that your company identified the rail for repair 21 months before the crash.
E: Yes. We noted the problem.
I: So the faulty rail was discovered 21 months before the crash. But nothing was done?
E: Not exactly –
I: Why wasn't the job completed?
E: There was a backlog of essential maintenance work, which was waiting to be done. Also, this job should have been carried out by another contractor, who had agreed to do the work. But they didn't do it.
I: Was *anything* done?
E: Yes, they delivered a replacement rail and left it alongside the track adjacent to the faulty rail.
I: And how long was the replacement rail lying there?
E: I believe it was there for six months.
I: Six months! So if the company that was responsible for the maintenance hadn't delayed for so long, the accident wouldn't have happened?
E: That's right. The faulty rail caused the train crash. It should have been repaired, but it wasn't.
I: Let's turn now to gauge corner cracking. What has been the policy for dealing with this in the past?
E: In the past, the policy has been to replace the whole rail where necessary.
I: Are you aware of the preventative maintenance known as preventative grinding at regular intervals?
E: Yes, I know about this system. They use it in the USA and in Sweden.
I: But you haven't considered using it?
E: No, the grinding machines are very expensive.
I: Is it more expensive to replace defective rails or to carry out preventative measures like grinding?
E: I don't know. I don't have the figures.
I: However, I understand that since the crash, a decision has been made to invest in three grinding machines, costing a total of £6 million.
E: Yes, I've read that, too, though I wasn't part of the decision-making process ...

27

I: Another matter, now. Records show that your company identified the rail for repair 21 months before the crash.
E: Yes, we noted the problem.
I: So the faulty rail was discovered 21 months before the crash. But nothing was done?
E: Not exactly –
I: Why wasn't the job completed?
E: There was a backlog of essential maintenance work, which was waiting to be done. Also, this job should have been carried out by another contractor, who had agreed to do the work. But they didn't do it.
I: Was *anything* done?
E: Yes, they delivered a replacement rail and left it alongside the track adjacent to the faulty rail.
I: And how long was the replacement rail lying there?
E: I believe it was there for six months.
I: Six months! So if the company that was responsible for the maintenance hadn't delayed for so long, the accident wouldn't have happened.
E: That's right. The faulty rail caused the train crash. It should have been repaired, but it wasn't.

catastrophic
compressive
excessive
faulty
inadequate
non-destructive
significant
temporary
tensile
undersized

Unit 11 Materials

A: Our sales of skis have been very successful in the past year. As we have some cash reserves now, I would suggest that we turn our attention to the manufacture of snowboards. However, before we do anything else, we could carry out some specific market research into local, national and international markets.
B: We don't have a product range yet, not even a prototype. Why don't we start an R&D programme that could run concurrently? Then we should end up with a product specification and a marketing plan at about the same time. What about hiring an R&D specialist with expertise in manufacturing snowboards?
A: That's possible. In fact, that's a very good suggestion. Let's try finding someone with the right kind of expertise. Would you like to take charge of that, to start with?
B: OK, I'll get onto that straight away.
A: And let's meet again in a week's time to check on progress.

I've been doing some research into the manufacture of snowboards and this is what I've found out. A snowboard is a laminated board, that is a board made up of several layers. The top of the board is a layer of acrylic, with the manufacturer's logo and artwork, which are very important. Then

come several layers of fibreglass, above and below the hardwood core. That goes onto the base.
The core of the board is hardwood because hardwood has lots of good properties. It has long fibres. That means reduced vibration, so the board shakes less as it goes over the snow. Secondly, hardwood is strong. And it's also rigid in torsion. So torsional rigidity, OK?
Above and below the wooden core you put layers of laminated fibreglass. Two layers above and two layers below, making four in total. That's kind of standard. The fibreglass is light, stiff and it's strong in torsion.
Some manufacturers add extra layers, for example, carbon fibre, Kevlar®, or honeycomb aluminium. Kevlar®'s popular and I think that's what we should go for. It provides elasticity and strength.
The base is made of polyethylene plastic. That's the same for all kinds of boards. Now, there are two kinds of base; one is extruded and the other is sintered, but I recommend extruded. Basically, it's a cheaper, low-maintenance board. It's smoother than the sintered board, but it's slower, and it doesn't soak up wax. That means you can't put wax on an extruded base to make it go faster over the snow. So an extruded board ends up with more surface friction than the sintered board. But on the other hand, it's cheaper and it's easier to repair. So the characteristics of the sintered base are the exact opposite of the extruded base.
I think that's about all. To finish, I've got a video clip off the internet for you. It shows the kind of machinery you need if you decide to manufacture …

absorbent, absorbency
ductile, ductility
durable, durability
elastic, elasticity
flammable, flammability
flexible, flexibility
malleable, malleability
non-flammable, non-flammability
plastic, plasticity
rigid, rigidity

Unit 12 Opportunities

A: Let's get started, then. There's a lot to discuss. Peter, you've got a list of four technologies that cars can use.
B: Well, five actually, if you look at the report in front of you. First, solar power. That's an existing technology, which is proven. But there will be technological improvements in the future.
A: Such as?
B: The Chinese are developing lighter panels and also the cost is coming down at a rate of about 5% per annum.
A: That's good news. So companies are at the prototype stage. Has any company started mass production?
B: No, not yet. Next, battery exchange, which takes place at switch stations.
A: Can you remind me?
B: Yes. You drive into a service station with a nearly flat battery. Your battery is removed and an exchange battery, fully charged, is placed in your car.
A: OK, so a few things to sort out, but possible.
B: Again, the cost of the system is likely to fall, and new generations of batteries will provide longer ranges and reduced weight. But infrastructure will be needed. And this is a chicken-and-egg situation.
A: Sorry? Chicken-and-egg? What's that?
B: The government won't subsidise investment in infrastructure – that is the provision of switch stations – until there are enough electric cars to use it, and …
A: And people won't buy electric cars until the infrastructure is in place. Of course. And companies won't want to make them, either.
B: No, but one switch station scheme is already in operation in Denmark, and more are planned. Now, the third system on my list is moving ahead rapidly. That's domestic battery recharging with plug-in systems. The research and development has been done, the prototypes have been tested, and one company is already in production.
A: So, it's up and running?
B: And running very well, thank you. The company is planning to produce 50,000 units next year and 200,000 units worldwide the year after. It's got a range of 160 kilometres, and a top speed of 140 kilometres per hour. A small recharging infrastructure exists, but only in a few countries.
A: Well, that's better than nothing. Anything new and fancy, like non-contact recharging?
B: No, it's simple plug-in recharging. But listen to this: they've got the recharging speed down. It charges to 80% of capacity in under 30 minutes.
A: I'm impressed!
B: Next system, the hydrogen fuel cell car. Now this system hasn't made much progress since we last discussed it. The R&D has been done and prototypes exist, but nobody has any plans to manufacture.
A: And of course the same problems with putting an infrastructure in place.
B: Exactly. That won't happen for a long time, that's my guess.
A: So, what's the fifth system, then?
B: It's the hybrid petrol-electric engine that we all know about. All the big manufacturers have got a hybrid in their range, and the number of different hybrid models on sale is expected to rise from 11 to 52 very soon.
A: 52 hybrid models to choose from! That's a huge change.
B: And the market share of hybrids is predicted to rise from 2.5% to 4.2% in the next two years.
A: So, everything's in place for servicing and repairs then. But it's not a zero-emission car, is it? It still pumps out emissions.
B: But it's a low-emission car. And until we've got good zero emission systems up and running and have got the infrastructure in place to keep them running, we're going to need those hybrids for a few years longer. It's intermediate low-emissions technology, but it works.

A: Peter, will you give us a presentation of your predictions for the short-term, medium-term and long-term use of these cars, please?
B: Sure. If you look at this slide, you'll see I've got predictions for three different types of economy, depending on the types of energy that are readily available in that zone. First, we have solar power, which is the energy of choice in the tropics, of course. Then in the next row down, we have countries with an abundant supply of cheap electricity generated by hydro-electric or geothermal schemes, or a combination. These are countries like Switzerland, Norway and Iceland. And of course, there are African countries with big hydro-electric schemes, too. Finally, we have countries with a developed nuclear power industry, which currently include France, the USA and Russia.
In the tropics, we expect that solar power will continue as the primary system in the short, medium and long term. It is, of course, the cheapest system to run, and it's the best system for the environment, too!
In the second group of countries, there is likely to be a rapid switch in the short-term to battery recharging, which is the least expensive system to buy. This will initially be with domestic plug-in systems, but we may get non-contact systems, too. This switch will take place because no additional infrastructure is required. However, in the medium and long term, we can expect a switch to battery exchange systems as the infrastructure required for this is built up and extended. It's probably the most convenient system for car drivers, as well.
Finally, in the nuclear countries I expect a similar pattern, in other words, a rapid switch in the short-term to battery recharging. Again, in the medium term, there is likely to be a switch to battery exchange systems. However, in the long term, there could be a switch to hydrogen fuel-cell systems, because of the type of energy output from nuclear power stations. This is not a firm prediction, but a possibility. But the major disadvantage of this system is that it requires the greatest investment in infrastructure.

Nouns (climate change)	Adjectives
atmosphere	convenient
concentration	exceptional
cyclone	high-performance
emission	huge
gigatonne	massive
glacier	renewable
ice cap	similar
pattern	spectacular
prediction	staggering
trend	technological
	tremendous

Answer key

1 Systems

1 Rescue

1 1 h 2 f 3 g 4 e 5 a 6 c 7 d 8 b

2 1 emergency beacon 2 coastguard 3 emergency signal 4 satellite 5 flares 6 inflate 7 life raft 8 winch

3 1 an emergency beacon 2 a sinking fishing boat 3 the fishermen's 4 a signal 5 a helicopter 6 one of the red flares 7 the fishermen 8 the coastguard station

2 Transmission

1 37.5 kilohertz 14,000 feet once per second 30 days 6 years in a container of water cylindrical orange

2 1 fastened 2 detached 3 activates 4 automatically 5 transmit 6 float 7 winched up

3
1 … TWA Flight 800, which crashed …
2 … in a fuel tank, which exploded.
3 … at Boston, where the weather was fine.
4 … by another pilot, who was flying …
5 … at Boston airport, from where he contacted …
6 … in the area, which lies off the coast …
7 … by divers, who were guided to it …
8 … to the shore, from where it was taken away …

3 Operation

1 1 Place C 2 Fasten F 3 inflate A 4 Remove H 5 Release G 6 Ensure D 7 Pull, push B 8 slide E

2 1 8 people 2 automatic 3 in vinyl bag 4 can turn upside down 5 can right itself

3
1 **inflated** canopy; **insulated** floor and canopy; water-**activated** lights on canopy; system for collecting **rainwater**
2 inflation triggered **automatically**; **two** buoyancy chambers inflated, to make the **walls**; inflation forces **open** the carrying bag; inflation time: **12 seconds**
3 **two** chambers fill with water; upper chamber fills through portholes in the chamber **walls**; **lower** chamber fills through a **one-way** valve, which **lets** water in, but not out.
4 waves of > **ten** metres, winds of > **300** kph; life raft self-**rights**

4 Word list

1 1 satellite signal 2 low-altitude orbit 3 radio frequency 4 operating range 5 air-sea rescue 6 safety device 7 rescue team

2 1 emergency beacon 2 radio signal 3 ground station 4 satellite signals 5 national centre 6 rescue centre 7 rescue team

3 1 antenna 2 HRU 3 beacon 4 lever arm 5 magnet 6 flare

2 Processes

1 Future shapes

1 1 composite 2 aerospace 3 carbon fibre 4 fuselage 5 aircraft 6 plastic 7 deck 8 report 9 construct 10 bridge 11 engineer
Vertical word: manufacture

2
1 The government will definitely cancel the manned space exploration programme.
2 It's likely that they will provide more money for robotic exploration of the solar system.
3 Scientists will probably not / probably won't develop new heavy-lift rockets in the near future.
4 They will certainly extend the life of the International Space Station beyond 2020.
5 They will probably ask commercial firms to play a bigger part in future.
6 Space travel to low-Earth orbit will possibly become more affordable.

2 Solid shapes

1 1 f 2 c 3 a 4 b 5 e 6 g 7 d

2 1 E 2 C 3 D 4 B 5 A
1 uses 2 is used 3 is loaded 4 is heated 5 rotates 6 melts 7 coats 8 is coated / has been coated 9 is cooled 10 cools / is cooled 11 shrinks 12 is removed

3 Hollow shapes

1
-ing: blowing, casting, cooling, heating, melting
-ion: ejection, expansion, extrusion, inflation, rotation
-ment: movement
-er: roller
-ure: closure
no change: transfer

2 air pump heater mould with holes thermoplastic sheet

3
1 heating
a) heater
b) thermoplastic sheet
2 positioning
c) mould with holes
3 stretching
4 vacuum forming
d) air pump

4 1 soft, hard 2 air-exhaust holes 3 below, above 4 thinner 5 vacuum

4 Word list

1 Suggested answers
1 The polymer pellets are transferred from the hopper to the cylinder.

2 The screw in the extrusion moulder is rotated by an electric motor.
3 The cold polymer pellets are propelled along the cylinder.
4 The polymer pellets are heated and melted by heaters.
5 The warm, soft molten polymer is propelled along the cylinder.
6 The molten polymer is extruded into a mould by the machine.
7 The two halves of the mould are closed with the molten polymer inside.
8 The molten polymer in the mould is inflated and expanded by compressed air.
9 The plastic bottle shape is cooled.
10 The plastic bottle is ejected from the open mould (by the machine).

2 close, closure; compress, compression; eject, ejection; expand, expansion; extrude, extrusion; heat, heater; inflate, inflation; inject, injection; mould, mould; rotate, rotation; transfer [verb], transfer [noun]

3 1 hopper 2 screw 3 mould 4 electric motor 5 cylinder 6 polymer pellets

Review Unit A

Section 1

1 1 crashed 2 It 3 raced 4 where 5 went 6 These/They 7 which 8 which 9 fell 10 took 11 hit 12 its/the 13 burst 14 flew 15 ignited

2 1 k 2 e 3 f 4 a 5 l 6 i 7 c 8 j 9 b 10 g 11 h 12 d

3
1 completes
2 captures
3 avoid
4 studies
5 discovers
6 are transmitted
7 are received
8 (are) converted

Section 2

1 and 2

A Injection moulding: 3 12 6
B Pressure-die casting: 10 4 7
C Metal-rolling: 1 11 8
D Blow moulding: 5 9 2

3
1 SOFTEN
2 EXPANSION
3 CARBON FIBRE
4 ROLLER
5 PISTON
6 CHAMBER
7 EJECT
8 COMPONENT
9 PROPEL
10 COMPRESS
11 INJECT
12 POLYMER

3 Events

1 Conditions

1 1 have closed 2 Has there been 3 collided 4 has caught 5 did this happen 6 received 7 Have you discovered 8 have not had 9 have you drilled 10 We have drilled

2
1 would only undertake
2 could be
3 would the government want
4 did not know
5 would it choose
6 was/were
7 would take
8 would be
9 would be
10 struck
11 would only plan
12 was/were

3
1 What will happen if the weather is bad?
The pilot will cancel the take-off.
2 What will happen if the launching device fails to function?
The jet fighter will remain on the flight deck.
3 What will the pilot do if the jet engine fails after take-off?
The pilot will activate the ejection system.
4 What will happen if the pilot ejects after take-off?
His/Her parachute will open automatically.
5 What will happen if the pilot lands in the sea after ejection?
A helicopter will winch the pilot to safety.

2 Sequence (1)

1 1 up to a maximum altitude of 100,000 metres 2 if there is a problem with the launch 3 the abort engine thrusts the LAS (and the crew capsule) away from the rocket 4 by the attitude control engine 5 with explosive bolts 6 the jettison engine must fire 7 to allow the capsule to float down into the ocean 8 once the capsule has reached a safe altitude

2
1 Once
2 Then
3 When
4 As soon as
5 After
6 After
7 However
8 Now

3
1 After giving the order for re-entry, the pilot fires the thrusters and turns the shuttle tail first.
2 Once the shuttle reaches / has reached the upper atmosphere, the pilot fires the thrusters again and turns the shuttle nose first.
3 When the shuttle enters / has entered the upper atmosphere, hot gases surrounding the shuttle cause a radio blackout.
4 As soon as it fully re-enters / has fully re-entered the Earth's atmosphere, the shuttle is able to fly like an aircraft.
5 After picking up the radio beacon at the end of the runway, the pilot takes over control from the onboard computers.
6 After landing, the pilot deploys the parachute from the rear to slow the shuttle.
7 Once the shuttle lands / has landed, the crew follows the procedure to shut down the shuttle.
8 As soon as the crew leaves / has left the shuttle, the ground crew begin servicing it.

3 Sequence (2)

1
1 are stabilised
2 deploying
3 orient
4 ejects
5 detonated
6 propelled
7 restraining
8 jettisoned

2 1 b 2 g 3 d 4 e 5 c 6 f 7 h 8 a

3
Ejection possible at speed: zero – **1,400** kph
Ejection possible at altitude: zero – **82,000** feet
Stage 1 Ejection
Pilot **loses** radio contact
Pilot has emergency **oxygen** supply
Pilot **stays in seat**
Stage 2 Drogue parachute
Purpose: 1) **stabilises** and 2) **slows** the fall of the pilot
Released at **8,000** feet
Stage 3 Main parachute
Pilot is **lifted out of seat**
Stage 4 Landing

Radio signal sent from **personal locator beacon**
Activated **automatically**

4 Word list

1
1 oil rig
2 jet fighter
3 crew capsule
4 launch abort system
5 abort engine
6 attitude-control engine
7 jettison engine
8 ejection system

2 broke burnt, burned flew focused propelled sank spun struck took thrust

4 Careers

1 Engineer

1 Martha Bari 2 Italian 3 Robotico
4 Research & Development Technician
5 Developing new devices for artificial arms and legs
6 Degree in Mechanical Engineering; Masters in Bio-medical Engineering 7 University of Ottawa 8 Speaks French and Italian

2 1 1 am working 2 specialises 3 spend 4 am doing 5 complete 6 am going to have / am having 7 don't have

2 1 work 2 work 3 stay 4 am developing 5 are starting / are going to start 6 have 7 am going to start / am starting

2 Inventor

1 1 device 2 hand-held 3 roughly 4 barrel 5 pulse 6 bullet 7 recoil 8 recoil

2
1 How long is the line?
2 How thick is the line?
3 How strong is the line?
4 What shape is the projectile? / What is the shape of the projectile?
5 How far can it send the projectile?
6 What is the maximum recoil?

3
1 The line on the 230 model is longer than the one on the 75 model.
2 The line on the 230 model is less thick than the one on the 75 model.
3 The breaking strength of the 230 model's line is 2000 N, while on the 75 model it is 1500 N.
4 The 230 model has a cylindrical projectile, whereas the 75 model has a plastic ball.
5 The 75 model has a shorter range than the 230 model.
6 The 230 model has a greater recoil than the 75 model.

3 Interview

Surname / First name(s)	GALLINI, Laura
Position applied for	**Technician for New Product Development**
Dates	**2008** (not 2005) – **now/present**
Position	**Senior** Technician (was Junior Technician)
Responsibilities	**Quality control; product testing; building a prototype for a new product**
Name and address of employer	Horton Engineering, Cleveland
Type of business	**Robotics; medical engineering**
Dates	**2003–2005**
Position	Apprentice Engineer
Responsibilities	Machining, finishing and some **quality control**
Name and address of employer	**BAMC plc** (was Farley Marine, Long Creek)
Type of business	Marine engineering: manufacture of engines and pumps
Dates	**2005–2008**
Qualification	**Bachelor of Applied Sciences**
Subjects / Skills covered	**Physics and Mechanical Engineering**
Name of institution	Albany College of Engineering, Albany

2
1 Which job are you interested in?
2 How long have you worked there?
3 What are your responsibilities in your current job?
4 Exactly what kind of business is your company in?
5 How long did you work at Farley Marine?
6 What was your job description there?
7 How long did your course last?
8 Why do you want to leave Horton Engineering?

3 for: three weeks, seven years, a week, six months, a month, two hours, 20 minutes, five days, a long time
since: 2011, yesterday, last week, January, 8 o'clock, Monday, 12th May, lunchtime

4 Word list

1
1 machine gun
2 bullets
3 barrel
4 target
5 recoil
6 Gene guns
7 helium
8 pulse
9 genetic
10 membrane
11 cell
12 prototype
13 inaccurate
14 tissue
15 modified
16 accuracy

2 1st syllable stressed: benefit, competence, previous
2nd syllable stressed: ambition, apprentice, apprenticeship, certificate, diploma, technician
3rd syllable stressed: engineering, institution, interpersonal

Review Unit B

Section 1

1 Take-off: i d f c a Landing: e h j b g

2
1 catapults
2 shuttle
3 tow bar
4 vertical drop line
5 runway centre line
6 coloured lights
7 flight deck
8 tail hook
9 arresting wires

Section 2

1 1 apprentice technician 2 blog activity / page 3 business activity / degree / experience / plan / qualification 4 career plan 5 company car 6 curriculum vitae 7 engineering degree / experience / qualification 8 job description 9 leisure activity 10 work experience / plan

2 1 more confident 2 but 3 older 4 more

experienced 5 faster 6 greater 7 worked 8 for 9 has worked 10 for 11 has been 12 since 13 left 14 ago 15 hasn't worked 16 since 17 but 18 have you received 19 more positive 20 better 21 while/whereas 22 whereas/while 23 Have you checked 24 but 25 less impressive

5 Safety

1 Warnings

1 and 2

1 b, consult main dealer
2 e, check all doors are shut
3 a, top up with oil
4 d, deflate tyre
5 c, fill up with fuel

3
1 any corrective action, brake heavily, counter-steer to the left
2 Where exactly, what colour is it, warning lights panel, red
3 all the time, monitoring it, from time to time
4 exactly, the problem, vibration in the steering wheel
5 visual or an audible signal, loud audible warning, ignition

4 1 For example 2 In other words 3 I agree with that 4 By the way 5 Anyway 6 Alternatively 7 for instance 8 I don't think that's a good idea. 9 That sounds good. 10 So

2 Instructions

1 1 shoes 2 back-plate 3 axle tube 4 Friction linings 5 brake line 6 shoes 7 drum 8 rear drum brakes 9 cable 10 lever

2 Suggested answers
1 The oil and the filter need to be changed. The radiator needs to be refilled with anti-freeze. The windscreen needs to be replaced if it is badly cracked.
2 The air bags should be replaced after ten years.
3 The windscreen doesn't have to be replaced if it's in good condition.
4 You need to check the screen wash system. You need to adjust the headlamps (if you are going to a country where people drive on a different side of the road).
5 You should check the oil level and top up if necessary. You should check the tyres, both the pressure and the tread depth. You should top up the screen wash. You should check all the lights (side lights, headlights and indicators).
6 These days, with improved technology, you don't have to check the battery or top up the radiator.

3 Rules

1 1 Hawk 2 (Reims-)Cessna

2
1 False. Air traffic control procedures at the airfield have already been modified.
2 True.
3 False. The danger of a collision was caused by several factors: 1) the air traffic controller was unable to handle the large workload 2) the breakdown in the monitoring of flight levels 3) the breakdown of radio communications.
4 False. The air traffic controller did not intervene to warn the pilots of the presence of other aircraft.
5 False. New staffing rotas are likely to reduce the danger of near misses at the airfield in future.

3
1 Pilots who are unfamiliar with the layout of this airport must order a 'Follow-Me' car before they land.
2 Do not ask for priority landing unless you have mechanical problems, a medical emergency or are dangerously short of fuel.
3 Only order mobility buggies and wheelchairs after you have landed.
4 Pilots must not shut down the engines until the ground staff have indicated that the aircraft is correctly positioned at the gate.
5 Do not embark onto your plane before filing a flight plan, with details of your route and destination.
6 Only take off after ensuring that you have sufficient fuel for the flight + 10%.
7 In snowy weather, aircraft must not attempt take-off without de-icing in the holding area.
8 Only move onto the runway if you have received permission from the control tower.

4 World list

1
1 cable
2 handbrake
3 steering wheel
4 brake pedal
5 disc brake
6 brake line
7 indicator
8 reservoir
9 brake line
10 master cylinder

6 Planning

1 Schedules

1
a) 0.5 °C
b) already
c) 2–3 °C
d) 2060
e) +5 °C
f) 2060
g) 0%
h) 2030
i) 1–3%
j) 2030
k) 60%
l) 2050
m) 30%
n) 2020
o) 60%
p) 2050

2
1 you think, will push up
2 could happen
3 could we avoid
4 your views
5 will that reduce, in the long term

3
1 I don't agree with you at all. I just can't go along with that.
2 You have a point / a good point there. I'm happy with that.
3 I'm not sure about that. Let's think again about it.
4 That sounds all right / about right. Yes, that's fine by me.

2 Causes

1
1 Phosphorus (used in the manufacture of detergents) is stored in water **owing to the reaction of phosphorus** with air.
2 The reaction happens **due to the addition of calcium carbonate**, together with water, to the gas.
3 The gas is purified **as a result of the complete removal of** polluting particles from the gas by the collection plates.
4 Emissions from power plants are lower with clean coal technology **as a result of the total purification of** the coal before it is burnt.
5 Port installation costs are high **owing to the liquefaction** of the LNG at a temperature of –1,620 Celsius before it is loaded onboard.
6 Greater crop yields are ensured **as a result of the automatic humidification of** the growing sheds.

7 The emergency rescue teams had problems **caused by the rapid solidification of** the volcanic ash after cooling.
8 The coal used in the power plant is a fine dust **due to its pulverisation** in the coal mill nearby.

2 1 particles 2 sulphur 3 electrode 4 chemicals 5 saline 6 desulphurise 7 aquifer 8 electrostatic 9 flaw 10 methane 11 ionise 12 impurities 13 gasification
Vertical word: precipitators

3 Systems

1

1 f	2 j	3 i	4 a
5 c	6 e	7 h	8 d
9 b	10 g		

2
1 Because hot water can be extracted from the underground reservoirs.
2 It comes from 13 production wells at Nesjavellir.
3 It travels in a transmission pipe (27 km long with a diameter of 90 cm).
4 It contains a mineral-rich saline solution which blocks the pipes. (So instead, heat exchangers are used to heat cold water, which is then distributed.)
5 Most of it (75.4%) comes from hydro-electric power. (24.5% comes from geothermal energy, and fossil fuels provide 0.1%.)

4 Word list

1

1 e	2 d	3 b	4 g
5 a	6 c	7 h	8 f

2 1 emitting, converted 2 pulverise, purify 3 is condensed 4 be liquefied, converted

Review Unit C

Section 1

1

1 d, q	2 i, o	3 f, k	4 h, p
5 b, r	6 e, j	7 c, n	8 g, m
9 a, l			

2

A 5	B 3	C 8	D 7
E 1	F 6	G 4	H 2

3 Suggested answers
1 Staff must not enter the building.
2 Doors must be locked at night.
3 Fork-lift trucks must not be driven in this part of the warehouse.
4 Workers must not eat or drink in this area.
5 All visitors must report to reception.
6 Pedestrians must not walk in the red zone.
7 Machine guards must be used at all times.
8 Ear protection must be worn when machinery is in operation.

Section 2

1

1 of, about	2 of	3 on to	4 through
5 over to	6 into	7 by	8 for

2 Model answer
Hi Bullent,
Many thanks for your email asking about incinerators. I'll try to answer your questions. Give me a ring, or email me, if you need more information.
1 Waste is very bulky. Fortunately, incineration reduces the mass of waste by 80–85%.
2 Incineration converts waste materials into heat, which can be used for space heating as well as electricity generation.
3 Of course, people are worried about pollution, but emission control is much better these days, and pollutants in the flue gas is reduced by particle filtration, using electrostatic precipitators. In addition, heavy metals such as mercury, lead and chromium are removed by acid gas scrubbers. By the way, the bottom ash from the incinerator is non-hazardous.
4 Incineration is a continuous process. In a moving grate incinerator, waste enters at one end and ash leaves at the other end. The heat is transformed into steam, which drives the turbine. The flue gases are cooled and then pass through the flue gas cleaning system.
5 This surprises some people, but there aren't the same concerns about pollution now that emission control is better. It is convenient for the reception of waste to site incinerators in cities. Also the heat can be utilised locally for space heating and electricity generation. Moreover, the heat can be used for increased electricity generation, for air-conditioning in summer or space heating in winter.
I hope this helps.
Best,
Karl

7 Reports

1 Statements

1 1 to the official, NO WORD 2 me 3 the clerk 4 to me, NO WORD 5 to him, NO WORD 6 him, NO WORD 7 to me, NO WORD 8 them

2
1 tell, inform, assure
2 say, report, explain, confirm
3 promise

3
1 'The airport has introduced a new scanning system.'
2 'The device can / is able to produce X-ray images of plaster casts.'
3 'It will be installed at all airports in the next six months.'
4 'Criminals have been able to transport knives inside plaster casts.'
5 'Conceal a knife inside the/this demonstration cast.'
6 'The body part being examined was/is placed next to the device, not in it.'
7 'The device uses a new system called 'backscatter imaging'.'
8 'Backscatter imaging can be used safely on all passengers.'
9 'The first group of operators is already undergoing training.'
10 'Read the user manual before tomorrow.'

2 Incidents

1 1 detection zone 2 magnetic field 3 reflected pulse 4 direct current 5 coil 6 audio circuit 7 resistor 8 pulse-induction technology

2
1 False. The hand-held remote display screen can operate at a distance of 65 m from the metal detector.
2 False. The display screen shows where a metal object is located on the passenger.
3 False. The WTMD is based upon eight separate coils of wire, arranged vertically.
4 True.
5 True.
6 False. A reflected pulse disappears more slowly after a metal object is detected.
7 True.

8 False. The audio output sounds higher and louder as the metal object approaches the coil.

3 1 As, While 2 while 3 while, as
4 While, When 5 when 6 While

4 1 While the passengers were checking in, the ground crew was refuelling the Airbus.
2 While the ground crew was loading the luggage into the Airbus, it started to snow.
3 A 747 landed while the ground crew was de-icing the Airbus.
4 The 747 slid off the runway while it was taxiing towards the terminal building.
5 While the Airbus passengers were waiting in the departure lounge, they heard that their flight was cancelled.
6 While the Airbus passengers were asking for information, the 747 passengers were waiting to leave their aircraft.

3 Progress

1 1 B 2 H 3 F 4 C
5 A 6 E 7 G 8 D

2 1 How are you getting on?
2 What systems have you come up with?
3 What about that other system that we spoke about?
4 Have you looked into that?
5 Yes, I see what you mean.
6 Go on, tell me more.
7 So how does the system operate?
8 I'll catch up with you after the weekend.

3 1 PIN numbers 2 scanning 3 video camera technology 4 8 and 35 cm 5 digital
6 converted 7 database 8 activated
9 iris identity 10 security door

4 Word list

1 1 capacitor 5 capacitance
2 conductor plates 6 discharge circuit
3 dielectric 7 capacitor
4 capacitor plates 8 resistor

2 amplifier capacitance capacitor conductor dielectric magnetism microchip resistor sensor terminal

8 Projects

1 Spar

1 1 mooring line 5 tank farm
2 tug boat (for towing tankers) 6 pipeline network
3 seabed 7 power plant
4 crude oil pipeline 8 pumping station

2 Suggested answers
1 A Have the foundations for the pumping station been laid yet?
B Yes, they were laid in January, four weeks ago.
A What about the pumps? Have they been installed?
B No. That job has already been started, but it hasn't been finished yet.
2 A Have the site roads been constructed yet?
B Yes, they've already been constructed. They were constructed three weeks ago.
A What about the pipeline ring main? Has it been installed?
B Not completely. That job has already been started, but it hasn't been finished yet.
3 A Have the test boreholes been drilled yet?
B Yes, they were drilled in January, five weeks ago.
A What about the foundations for the water storage tanks? Have they been laid?
B Yes, that job was finished two weeks ago.
4 A Have the tank sections for the water storage tanks been ordered yet?
B Yes, they've already been ordered. They were ordered in January, four weeks ago.
A Have they been assembled yet?
B No, that job hasn't been started yet.
5 A Have the fire guns been delivered yet?
B Yes, they were delivered one week ago.
A Have they been used in training?
B No, not yet. Training is planned for next week, and for two more weeks in March.

2 Platform

1 All the questions are answered, except 4 and 7.

2 1 yes 2 no 3 2
4 (not given) 5 yes 6 yes
7 (not given) 8 no

3 a) 103 million b) 32 million c) 28 million
d) 462,300 e) 1993 f) 20,000 g) 1,084 h) 660

4 1 to allow 5 to pass
2 by constructing 6 by means of
3 by dumping 7 to allow
4 by means of 8 to provide

3 Drilling

1 1 f 2 a 3 e 4 c
5 b 6 d 7 g
8 k 9 m 10 h 11 n
12 i 13 l 14 j

2 1 was laid 2 was driven 3 was taken
4 was awarded 5 were drilled 6 was pushed
7 was missed 8 was enlarged 9 were chosen
10 were pulled, were rotated

4 Word list

1 1 derrick 7 hook
2 oil rig 8 swivel
3 drilling platform 9 kelly
4 crown block 10 turntable
5 winch 11 drill pipe
6 travelling block 12 drill collar

2 First syllable stressed: cylinder drilling pipeline stabilise travelling turntable
Second syllable stressed: continuously directional hydraulic obstruction stability sufficient

Review Unit D

Section 1

1 Suggested answers
1 She informed him that he had to buy three seats together because of his broken leg.
2 The steward told the passenger not to smoke in the departure lounge.
3 The helpline confirmed that we could carry 10 kg of medical equipment free of charge.
4 The airline official assured her that the plane would not leave without her.
5 The check-in clerk promised he would book us on the next possible flight.

6 The check-in clerk explained to him that folding wheelchairs were carried free of charge.
7 The steward instructed the children not to sit in the emergency exit row.
8 The pilot ordered the crew to evacuate the aircraft immediately.

2
1 permanent magnet
2 magnetic poles
3 electromagnet
4 armature
5 coil
6 slip rings
7 brushes
8 external power circuit
9 shaft
10 north pole
11 south pole
12 commutator

Section 2

1 1 swivel 2 kelly 3 casing
4 drill string 5 drill collar 6 drill bit

2
A rock cuttings
B reserve pit
C mud return line
D shaker
E hose
F mud pump

3
A injection moulding
B power plant
C chemical reaction
D reflected pulse
E collection plate
F brake pad
G magnetic field
H well head
I rock face
J lever arm
K protective sleeve
L insulating jacket
M wind sock
N brake shoe

9 Design

1 Inventions

1 Suggested answers
1 The GT Sport accelerates a great deal faster than the Roadster.
2 The Roadster is a lot more economical than the GT Sport.
3 The GT Sport is a little more stable than the Roadster.
4 The Roadster's suspension is slightly better than the GT Sport's.
5 The maximum speed of the GT Sport is 71 kph faster than the Roadster's.
6 The Roadster's engine is two thirds as powerful as the GT Sport's.
7 The GT Sport's storage space is twice as large as the Roadster's.
8 The GT Sport is 50% more expensive than the Roadster.

2 Suggested answers
1 The Roadster's acceleration is much worse than the GT Sport's.
2 The Roadster uses two thirds as much fuel as the GT Sport.
3 The Roadster is slightly more unstable than the GT Sport.
4 The GT Sport's suspension is a little less comfortable than the Roadster's.
5 The Roadster's top speed is two thirds as fast as the GT Sport's.
6 The GT Sport's engine is one and a half times more powerful than the Roadster's.
7 The Roadster has half as much storage space as the GT Sport.
8 The Roadster is 30% cheaper than the GT Sport.

3 Model answers
1 At weekends, I use the bus much more often. On weekdays, I use the train twice as often as the bus.
2 The car is much more comfortable than either the bus or the train.
Buses are a great deal more crowded in the rush hour than trains.
For me, the car is far more convenient than the train and a little more convenient than the bus.
Parking a car is less convenient at busy times than getting off a bus.
A train ticket is five times more expensive than a bus ticket for a similar bus journey.

2 Buildings

1 1 C 2 D 3 B 4 A

2
Burj Khalifa
1 Dubai 2 United Arab Emirates 3 828 m
4 50 m 5 206 6 464,000 sq m 7 Y-shaped
8 hotel, apartments, restaurants, outdoor observation deck 9 25,000 windows

Taipei 101
1 Taipei
2 Taiwan
3 509 m
4 80 m
5 101 + 5 = 106
6 412,000 sq m
7 8 tapered sections of 8 floors each, on square, tapered base
8 offices, conference centres, retail shops, restaurants
9 designed to withstand earthquakes and strong winds

Shanghai World Financial Centre
1 Shanghai
2 China
3 494 m
4 79
5 101
6 381,600 sq m
7 square at the base, tapers to rectangular top
8 hotel, offices, conference centres, shopping malls
9 public observation deck on 100th floor

3 Suggested answers
3 Burj Khalifa is by far the tallest building of the three.
3 Shanghai World Financial Centre is the shortest building of the three.
4 Taipei 101 has the deepest foundations.
4 Burj Khalifa has the least deep foundations of the three buildings.
5 Burj Khalifa has by far the greatest total number of storeys.
5 Shanghai World Financial Centre has the least number of storeys.
6 Burj Khalifa has by far the greatest floor area.
6 Shanghai World Financial Centre has the least extensive floor area of the three buildings.

3 Sites

1
1 Picture F; elevation, perpendicular, zigzag
2 Picture D; rectangular, vertical, diagrid
3 Picture B; wings, storeys, taper
4 Picture G; irregular, inclined, curved

2 Model answers
A
The department building has a semi-circular floor plan, with the circular front elevation made of glass and stone. The building rises in three tapering storeys, with metal roofs. The top storey looks like an observation lounge.
C
The student hostel is constructed of white concrete and glass. Some solid vertical concrete columns are placed at intervals along the front elevation. All the windows are either square or rectangular.

E
The assembly room is a three-storey six sided building. The windows on the ground and first floor are rectangular, surrounded by elevations of natural stone. The third storey is a roof terrace. A six-sided stone framework surrounds a six-sided conical pyramid constructed of glass with aluminum strips. The assembly room is linked to the main building by enclosed walkways on two levels.
H
The research centre has a rectangular floor plan. It has a flat metal roof, supported at the edges by perpendicular steel columns. The front and side elevations are constructed of glass and brick. On the left you can see some irregular conical roofs, which cover the bicycle racks.

4 Word list

1
1 pointed
2 doughnut-shaped
3 inclined
4 zigzag
5 perpendicular
6 circular
7 elliptical
8 curved
9 tapered
10 diagonal

2 1 optional 2 extensive 3 compact
4 stable 5 pointed

3

Nouns (buildings)	Adjectives (shapes)
aluminium	circular
architect	curved [one syllable]
beauty	curvilinear
concrete	diagonal
elevation	diagrid
functionality	doughnut-shaped
innovation	elliptical
storey	inclined
	oval
	perpendicular
	pointed
	semi-circular
	tapered
	zigzag

10 Disasters

1 Speculation

1 1 compression 2 impact 3 wear
4 fracture 5 buckling 6 thermal shock
7 investigation 8 metal fatigue 9 corrosion
10 collapse 11 tension
Vertical word: speculation

2
1 The rail crash must have been caused by a broken rail.
2 The rail may have suffered thermal shock due to excessive heat.
3 The aircraft might not have had a mechanical failure.
4 The flight crew could have fallen asleep due to cabin depressurisation and lack of oxygen.
5 The captain might have wanted to save time by steering close to the headland.
6 He may not have realised that the depth of water in the channel was insufficient.
7 The houses can't have collapsed in the earthquake because of a design error.
8 The disintegration of the bridge might not have been caused by substandard concrete.

2 Investigation

1 1 No 2 No 3 No 4 21 months before crash
5 other company 6 backlog of essential maintenance work 7 6 months 8 rail replacement 9 USA, Sweden 10 3 11 £6 million

2
1 should have been
2 hadn't
3 wouldn't have
4 should have been

3 Reports

1
1 should have sounded
2 must have failed
3 must have been
4 should place
5 should be checked
6 should be installed
7 should inspect
8 should be increased

2
Introduction: f, i
Background: c, l
Method: b, j
Findings: a, e
Conclusions: d, h
Recommendations: g, k

4 Word list

1

NOUNS	VERBS	ADJECTIVES
catastrophe		catastrophic
compression	compress	compressive
excess	exceed	excessive
destruction	destroy	destructive
buckling	buckle	
collapse	collapse	
corrosion	corrode	corrosive
disintegration	disintegrate	
fracture	fracture	
rust	rust	rusty

2
1 d 2 h 3 f 4 b
5 e 6 g 7 c 8 a

3 catastrophic compressive excessive
faulty inadequate non-destructive
significant temporary tensile undersized

Review Unit E

Section 1

1 Suggested answers
1 The AX 210 lamp has a far higher wattage than the RS 90.
2 The lamp lumen output of the AX 210 is much greater than the lumen output of the RS 90.
3 The RS 90 weighs a great deal less than the AX 210.
4 The AX 210 is a lot longer than the RS 90.
5 The beam of the RS 90 is slightly narrower than the beam of the AX 210.
6 The RS 90 has a little more control over the aiming point than the AX 210.

2 Suggested answers
1 The wattage of the AX 210 is exactly twice as powerful as the wattage of the RS 90.
2 The lumen output of the KH 240 is approximately 10% greater than the lumen output of the AX 210.
3 The RS 90 is roughly half as heavy as the KH 240.
4 The AX 210 is 100 mm wider than the KH 240.

5 The beam of the KH 240 is about ten times wider than the beam of the RS 90.
6 The RS 90 has five times as much control over the aiming point as the KH 240.

3 Suggested answers
1 The RS 90 has by far the least powerful wattage.
2 The KH 240 has easily the greatest lumen output.
3 The RS 90 is by far the lightest lamp of the three.
4 The RS 90 is easily the narrowest lamp of the three.
5 The KH 240 lamp has by far the widest beam.
6 The RS 90 lamp has easily the greatest control over the aiming point.

4 1 circular curved curvilinear doughnut-shaped elliptical oval
2 diagonal diagrid perpendicular pointed tapered zigzag
3 conical cylindrical semi-circular

Section 2

1

Introduction C	Findings F
Background E	Conclusions B
Method A	Recommendations D

2 1 The fire would not have happened if there had not been a build-up of flammable gases.
2 The accident would not have happened if Worker A had received clear instructions.
3 If the foreman had not gone on a meal break, he would have supervised Worker A.
4 If he had not obtained a suitable fitting, he would not have connected the airline to the ring main.
5 If Worker B had not gone above deck to get some fresh air, he would not have escaped the fire.

3 (a) Temporary maintenance and repair workers should have received training before starting work.
(b) Temporary maintenance and repair workers should have been properly supervised.
(c) All dockyard gas ring mains should have been securely labelled and colour-coded.
(d) Bulk oxygen should have been treated at the supplier's depot before delivery.
(e) Worker B should not have lit a cigarette.

11 Materials

1 Equipment

1 Suggested answers
1 Aramid fibre resists impact.
2 Polyurethane foam absorbs impact.
3 You can bend carbon fibre, but you can't break it.
4 You can bend and stretch rubber, but you can't break it.
5 Aramid fibre can be stretched a little, but it can't be broken.
6 TPU can be bent and stretched, but it can't be broken.
7 Nylon synthetic fibre stretches a little, but it doesn't break.
8 Wood burns and bends, but it doesn't stretch.

2 Model answer
Ali Said
Manager, Muscat Athletics Club

Dear Mr Said,

Thank you very much for inviting me to make a proposal to supply your club with running shoes.

At the presentation last week, I demonstrated our Plym running shoes and you then kindly invited me to send you a proposal.

As I explained, our running shoes are designed to give lightness, durability and flexibility. The shoes combine the strength and flexibility of thermoplastic polyurethane with the cushioning and shock absorbent qualities of foam-blown polyurethane. In addition, mesh is added to the upper part of the shoe, allowing the foot to breathe.

The shoes give increased support and elasticity around the heels and under the soles, allowing the runner to run long distances without tiring.

Details of all the materials used in the shoes can be found in the attachment to this letter.

My company proposes to supply these shoes at the unit price (per pair) of US$49.50. Package and delivery is free of charge, and delivery dates are a maximum of two weeks after receipt of order.
This offer is open for 28 days from your receipt of this proposal.

I look forward to hearing from you with a firm order in due course.

Yours sincerely,

....

2 Properties (1)

1 1 b 2 d 3 e 4 c 5 a

2

1 flammability	4 ductility
2 absorbency	5 rigidity
3 durability	6 malleability

3 Properties (2)

1 1 snowboards
2 a marketing plan
3 in a week's time

2 1 I would suggest that 2 We could
3 Why don't we 4 What about
5 Let's try 6 Let's

3 1 hardwood
a) long fibres b) strength c) torsional rigidity
2 4 layers
a) lightness b) stiffness c) torsional strength
3 Kevlar®
a) elasticity b) strength
4 extruded
smoother, slower, non-porous, greater surface friction, lower cost, easier to repair

4 Suggested answers
1 Extruded boards cannot absorb wax.
2 Fibreglass layers are able to withstand torsion and return to their original shape.
3 A waxed board has the capacity to reduce drag over all types of snow.
4 Sintered boards are incapable of reaching high speeds unless they are waxed.
5 The longer board has the capacity to reach faster speeds in downhill races.
6 A snowboarder has the ability to split the board into two mono-skis.

7 We are capable of selling 18,000 units per annum within three years.
8 However, we do not have the capability to achieve / of achieving substantial sales this year.

4 Word list

1 Suggested answers
1 waterproof
2 childproof
3 stain-resistant / heat-tolerant
4 bulletproof
5 ovenproof
6 impact-absorbent
7 corrosion-resistant / impact-resistant
8 water-resistant
9 fireproof
10 shock-resistant / waterproof
11 heat-tolerant / ovenprooof
12 impact-resistant

2
1 f
2 c
3 h
4 a
5 d
6 g
7 e
8 b

3 absorbent, absorbency
ductile, ductility
durable, durability
elastic, elasticity
flammable, flammability
flexible, flexibility
malleable, malleability
non-flammable, non-flammability
plastic, plasticity
rigid, rigidity

12 Opportunities

1 Threats

1
1 By 2040, CO_2 emissions will have risen to 18 gigatonnes per year.
2 By 2080, CO_2 emissions will have increased by 23 gigatonnes per year.
3 By 2080, CO_2 concentrations will have grown by 400 parts per million.
4 By 2100, CO_2 concentrations will have climbed to 920 ppm.
5 By 2080, CO_2 emissions will have stabilised at 6 gigatonnes per annum.
6 By 2100, CO_2 emissions will have dropped by 2 gigatonnes p.a.
7 By 2080, CO_2 concentrations will have increased to 520 parts per million.
8 By 2100, CO_2 concentrations will have risen by 200 ppm.

2 Suggested answers
1 By 2060, the glaciers south of the Himalayas will have melted. The great rivers coming from the glaciers will have run dry. All agriculture in the river valleys will have ceased. Entire populations will have been displaced.
2 By 2060, the ice caps of Greenland and the Arctic will have melted. The sea level will have risen by at least 0.4 metres. Low-lying areas will have (been) flooded, and some countries like the Maldives will have been abandoned.
3 By 2060, climate change will have caused extreme global warming. Many areas in the tropics as well as in the Mediterranean basin will have been affected by droughts and fires. Many populations in these areas will have been forced to migrate southwards and northwards to find more temperate climates.
4 By 2060, increased global temperatures will have been caused by continued burning of forested areas. By this time, parts of the frozen tundra of Siberia will have started to thaw, releasing methane into the atmosphere. Many forests of northeast Asia will have been destroyed by fires fuelled by methane.

2 Innovation

1
1 relies
2 similar
3 propels
4 drag
5 lift
6 friction
7 like
8 direction
9 inclined
10 keel
11 polycarbonate
12 apparent wind
13 Aerodynamic
14 rigid

2 Suggested answers
1 A helicopter doesn't resemble a glider in any way. It uses revolving blades to fly instead of wings. A helicopter is powered whereas a glider is unpowered.
2 A submarine propels itself under water, in the same way as a submersible does. But a submarine is manned, whereas a submersible is unmanned.
3 A drone resembles a glider, but a glider is piloted, whereas a drone is remote controlled.
4 A space shuttle is completely unlike a helicopter. The former uses wings instead of revolving blades and travels into space, whereas the latter flies at low altitudes only.
5 A jet ski is a little like a hovercraft; both can travel across water, but a jet ski can't travel across marshy ground.
6 A hovercraft resembles a helicopter in one way only. Both hover above the ground. However, a hovercraft hovers on a cushion of air instead of using revolving blades.

3 Priorities

1 and 2
1 A yes B yes C no D __ E 5% p.a.
2 A yes B yes C no D no E Denmark
3 A yes B yes C yes D yes E 200,000 units; 160 km; 140 kph; up to 80% in under 30 minutes
4 A yes B yes C no D no E __
5 hybrid petrol-electric engine A yes B yes C yes D yes E from 11 to 52; from 2.5 to 4.2%

3 A all: 1) solar power
B short-term: 3) battery recharging; medium- and long-term: 2) battery exchange
C short-term: 3) battery recharging; medium-term: 2) battery exchange; long-term: 4) hydrogen fuel-cell

4 1 solar power 2 solar power 3 battery recharging 4 battery exchange 5 battery exchange 6 hydrogen fuel-cell technology

4 Word list

1 1 in 2 NONE 3 on 4 NONE 5 into 6 onto

2
1 atmosphere
2 prediction
3 trend
4 pattern
5 exchange
6 weakness

3
1 emissions
2 gigatonnes
3 concentration
4 atmosphere
5 Patterns
6 glaciers
7 ice caps
8 cyclones
9 trends

Review Unit F

Section 1

1 Suggested answers

1. The first material is strong when it is pressed, and the second is strong when it is pulled.
2. The first material is rigid and doesn't break, and the second is soft and returns to its shape after an impact.
3. A waterproof watch is guaranteed to keep out all water, but a water-resistant watch may not be.
4. Flammable materials burn, and non-flammable materials don't burn.
5. The first material is strong when it is twisted and the second is strong when it is stretched.
6. Ductile materials can be pulled into a longer, thinner shape, and malleable materials can be rolled into a new shape.

2 1 able 2 can't 3 can 4 capabilities 5 ability 6 capable 7 unable 8 capacity 9 incapable

Section 2

1

1. will have been built, will have been added, will have been installed
2. will have been established, will have been issued
3. will have become, will have expanded
4. will have developed, will have changed

2

1. … is 25 metres long, instead of the usual 18 metres. Another difference is that it is a double-articulated truck …
2. The Archimedes water screw has exactly the same function as a traditional waterwheel. In the same way as … Unlike a traditional hydro-electric turbine, …
3. An unmanned drone, or remotely piloted aircraft, is very similar to … … instead of a pilot. … just as a pilot flies a plane, …
4. Laser drilling for oil is completely different from normal rotational drilling. … in contrast to a traditional derrick. Another difference is that the heat from the laser …